U.S.NRC
United States Nuclear Regulatory Commission
Protecting People and the Environment

NUREG-2109

I0482710

Technical Evaluation Report on the Content of the U.S. Department of Energy's Yucca Mountain Repository License Application

Administrative and Programmatic Volume

Office of Nuclear Material Safety and Safeguards

AVAILABILITY OF REFERENCE MATERIALS
IN NRC PUBLICATIONS

NRC Reference Material

As of November 1999, you may electronically access NUREG-series publications and other NRC records at NRC's Public Electronic Reading Room at http://www.nrc.gov/reading-rm.html.
Publicly released records include, to name a few, NUREG-series publications; *Federal Register* notices; applicant, licensee, and vendor documents and correspondence; NRC correspondence and internal memoranda; bulletins and information notices; inspection and investigative reports; licensee event reports; and Commission papers and their attachments.

NRC publications in the NUREG series, NRC regulations, and *Title 10, Energy*, in the Code of *Federal Regulations* may also be purchased from one of these two sources.
1. The Superintendent of Documents
   U.S. Government Printing Office
   Mail Stop SSOP
   Washington, DC 20402-0001
   Internet: bookstore.gpo.gov
   Telephone: 202-512-1800
   Fax: 202-512-2250
2. The National Technical Information Service
   Springfield, VA 22161-0002
   www.ntis.gov
   1-800-553-6847 or, locally, 703-605-6000

A single copy of each NRC draft report for comment is available free, to the extent of supply, upon written request as follows:
Address: U.S. Nuclear Regulatory Commission
           Office of Administration
           Publications Branch
           Washington, DC 20555-0001
E-mail: DISTRIBUTION.SERVICES@NRC.GOV
Facsimile: 301-415-2289

Some publications in the NUREG series that are posted at NRC's Web site address
http://www.nrc.gov/reading-rm/doc-collections/nuregs
are updated periodically and may differ from the last printed version. Although references to material found on a Web site bear the date the material was accessed, the material available on the date cited may subsequently be removed from the site.

Non-NRC Reference Material

Documents available from public and special technical libraries include all open literature items, such as books, journal articles, and transactions, *Federal Register* notices, Federal and State legislation, and congressional reports. Such documents as theses, dissertations, foreign reports and translations, and non-NRC conference proceedings may be purchased from their sponsoring organization.

Copies of industry codes and standards used in a substantive manner in the NRC regulatory process are maintained at—
           The NRC Technical Library
           Two White Flint North
           11545 Rockville Pike
           Rockville, MD 20852-2738

These standards are available in the library for reference use by the public. Codes and standards are usually copyrighted and may be purchased from the originating organization or, if they are American National Standards, from—
           American National Standards Institute
           11 West 42nd Street
           New York, NY 10036-8002
           www.ansi.org
           212-642-4900

Legally binding regulatory requirements are stated only in laws; NRC regulations; licenses, including technical specifications; or orders, not in NUREG-series publications. The views expressed in contractor-prepared publications in this series are not necessarily those of the NRC.

The NUREG series comprises (1) technical and administrative reports and books prepared by the staff (NUREG-XXXX) or agency contractors (NUREG/CR-XXXX), (2) proceedings of conferences (NUREG/CP-XXXX), (3) reports resulting from international agreements (NUREG/IA-XXXX), (4) brochures (NUREG/BR-XXXX), and (5) compilations of legal decisions and orders of the Commission and Atomic and Safety Licensing Boards and of Directors' decisions under Section 2.206 of NRC's regulations (NUREG-0750).

United States Nuclear Regulatory Commission

Protecting People and the Environment

NUREG-2109

Technical Evaluation Report on the Content of the U.S. Department of Energy's Yucca Mountain Repository License Application

Administrative and Programmatic Volume

Manuscript Completed: September 2011
Date Published: September 2011

Office of Nuclear Material Safety and Safeguards

ABSTRACT

This "Technical Evaluation Report on the Content of the U.S. Department of Energy's Yucca Mountain License Application; Administrative and Programmatic Volume" (TER Administrative and Programmatic Volume) presents information on the NRC staff's review of the U.S. Department of Energy (DOE) Safety Analysis Report (SAR), provided on June 3, 2008, as updated on February 19, 2009. The NRC staff also reviewed information DOE provided in response to NRC staff requests for additional information and other information that DOE provided related to the SAR. In particular, this report provides information on the NRC staff's evaluation of DOE's proposed administrative and programmatic activities regarding the following:

- Research and Development Program to resolve safety questions

- Performance Confirmation Program

- Quality Assurance Program

- Records, reports, tests, and inspections

- DOE organizational structure

- Key positions assigned responsibility for safety and operations

- Personnel qualifications and training

- Plans for startup activities and testing

- Plans for conduct of normal activities

- Emergency planning

- Controls to restrict access and regulate land uses

- Uses of geologic repository operations area for purposes other than disposal of radioactive wastes

CONTENTS

CONTENTS (continued)

CONTENTS (continued)

CONTENTS (continued)

EXECUTIVE SUMMARY

After docketing the U.S. Department of Energy (DOE) license application seeking a construction authorization for the proposed repository at Yucca Mountain, Nevada, the U.S. Nuclear Regulatory Commission (NRC) staff began documenting its review in a Safety Evaluation Report. On March 3, 2010, DOE filed a motion with the Atomic Safety and Licensing Board seeking to withdraw its license application to develop a repository at Yucca Mountain, Nevada. In June 2010, the Board denied the DOE motion. To date, petitions asking the Commission to reverse or uphold this decision are pending before the Commission.

On October 1, 2010, the NRC staff began orderly closure of its Yucca Mountain activities. As part of orderly closure, the NRC staff prepared this technical evaluation report (TER), a knowledge management document. This document captures the NRC staff's technical assessment of information presented in DOE's Safety Analysis Report (SAR), dated June 3, 2008, as amended, and supporting information. The TER describes the NRC staff's technical evaluation of the DOE SAR and, in particular, this TER Administrative and Programmatic Volume provides insights on administrative and programmatic activities in the context of geologic disposal. The TER was developed using the regulations at 10 CFR Part 63 and guidance in the Yucca Mountain Review Plan (YMRP). The TER does not, however, include conclusions as to whether or not DOE satisfies the Commission's regulations.

NRC's regulations at 10 CFR Part 63 provide site-specific criteria for geologic disposal at Yucca Mountain. These regulations prescribe requirements governing the licensing (including issuance of a construction authorization) of DOE to receive and possess source, special nuclear, and byproduct material at a geologic repository operations area (GROA) sited, constructed, or operated at Yucca Mountain, Nevada. Under 10 CFR Part 63, there are several stages in the licensing process: the site characterization stage, the construction stage, and a period of operations. The period of operations includes the time during which emplacement would occur, any subsequent period before permanent closure during which the emplaced wastes are retrievable, and permanent closure. In addition, the regulations at 10 CFR Part 63 represent a risk-informed, performance-based (RIPB) approach to the review of geological disposal. The RIPB approach uses risk information to focus the review to areas most significant to safety or performance. The TER includes discussions regarding how the staff used risk information in its review. For the TER Administrative and Programmatic Volume, risk information was used principally in the NRC staff's review of DOE's Performance Confirmation Program (Chapter 2 in this volume).

This technical evaluation report presents information on the NRC staff's assessment of the SAR DOE provided on June 3, 2008, as updated on February 19, 2009.[1] The NRC staff also reviewed information DOE provided in response to NRC staff's requests for additional information and other information that DOE provided related to the SAR. In conducting its review of DOE's SAR, the NRC staff was guided by the Yucca Mountain Review Plan.[2]

---

[1]DOE. 2009. DOE/RW–0573, "Yucca Mountain Repository License Application." Rev. 1. ML090700817. Las Vegas, Nevada: DOE, Office of Civilian Radioactive Waste Management.

[2]NRC. 2003. NUREG–1804, "Yucca Mountain Review Plan—Final Report." Rev. 2. Washington, DC: NRC.

DOE provided the following in its SAR:

- An identification of those structures, systems, and components (SSCs) of the geologic repository, both surface and subsurface, that require research and development to confirm the adequacy of design

- A description of the Performance Confirmation Program

- A description of the quality assurance program to be applied to the SSCs important to safety and to the engineered and natural barriers important to waste isolation

- A description of the program to be used to maintain the records

- The organizational structure of DOE as it pertains to construction and operation of the GROA, including a description of any delegations of authority and assignments of responsibilities, whether in the form of regulations, administrative directives, contract provisions, or otherwise

- Identification of key positions that are assigned responsibility for safety at and operation of the GROA

- Personnel qualifications and training program

- Plans for startup activities and startup testing

- Plans for conduct of normal activities, including maintenance, surveillance, and periodic testing of structures, systems, and components of the GROA

- A description of the plan for responding to, and recovering from, radiological emergencies that may occur at any time before permanent closure and decontamination or decontamination and dismantlement of surface facilities

- A description of the controls that DOE will apply to restrict access and to regulate land use at the Yucca Mountain site and adjacent areas, including a conceptual design of monuments that would be used to identify the site after permanent closure

- Plans for any uses of the GROA at the Yucca Mountain site for purposes other than disposal of radioactive wastes, with an analysis of the effects, if any, that such uses may have on the operation of the SSCs important to safety and the engineered and natural barriers important to waste isolation

A summary of the NRC staff review of these 12 areas is provided next.

Research and Development Program To Resolve Safety Questions

The SAR includes an identification of those SSCs of the geologic repository, both surface and subsurface, that require research and development to confirm the adequacy of design. For SSCs important to safety and for engineered and natural barriers important to waste isolation, DOE provided a description of a program designed to resolve safety questions, including a schedule indicating when these questions would be resolved. DOE did not identify any safety

x

questions and, therefore, did not describe a specific research and development program to address safety questions. However, DOE described how a research and development program, separate and distinct from the Performance Confirmation Program, would be developed and implemented should a safety question be identified in the future.

The NRC staff has reviewed the SAR and the other information submitted in support of the SAR. Because neither DOE in its SAR nor the NRC staff in its review of the SAR identified any safety questions, the NRC staff notes that a research and development program that is separate and distinct from DOE's Performance Confirmation Program is not needed at this time. DOE's description of its approach for developing and implementing a research and development program is consistent with the guidance in the YMRP and is reasonable as a general approach to use in addressing safety questions should a safety question be identified in the future.

Performance Confirmation Program

The SAR includes a description of the Performance Confirmation Program, which evaluates the adequacy of the supporting assumptions, data, and analyses in the SAR. DOE stated that key geotechnical and design parameters, including any interactions between natural and engineered systems and components, will be monitored and changes will be analyzed throughout site characterization, construction, emplacement, and operation to identify any significant changes in the conditions assumed in the SAR that may affect postclosure performance. DOE described its performance confirmation activities and stated it would provide to NRC, prior to test implementation, future performance confirmation test plans outlined in the SAR.

On the basis of the NRC staff's review of the SAR and other information submitted in support of the SAR, the NRC staff notes that DOE has provided a reasonable description of its Performance Confirmation Program that is consistent with the guidance in the YMRP.

Quality Assurance Program

The SAR includes a description of the quality assurance program that will be applied to the SSCs important to safety, and to the engineered and natural barriers important to waste isolation. DOE's quality assurance program is described in the Quality Assurance Requirements and Description (QARD).[3] DOE's QARD is organized consistent with the 18 Quality Assurance (QA) topics discussed in the YMRP. DOE stated the QARD applies throughout the proposed Yucca Mountain geologic repository design and construction activities. DOE stated that the QARD will be revised, as necessary, to address future activities related to facility operations, permanent closure of the repository, and decommissioning and dismantlement of the surface facilities.

On the basis of the NRC staff's review of the SAR and other information submitted in support of the SAR, the NRC staff notes that DOE provided a reasonable description of the quality assurance program, including the QARD, consistent with the guidance in the YMRP.

---

[3]DOE. 2008. DOE/RW-0333P, "Quality Assurance Requirements and Description (QARD)." Rev. 20. ML0801450334. Las Vegas, Nevada: DOE, Office of Civilian Radioactive Waste Management.

Records, Reports, Tests and Inspections

DOE described the recordkeeping and reporting programs for receipt, handling, and disposition of radioactive waste to provide a complete history of the movement of the waste from the shipper through all phases of storage and disposal. DOE described a program to maintain records of construction of the geologic repository operations area in a manner that ensures their usability for future generations. DOE also provided program descriptions for reporting deficiencies to NRC, performing tests for NRC or allowing NRC to perform tests, and allowing the NRC personnel access to the GROA and adjacent areas, and to DOE records, upon reasonable notice.

On the basis of the NRC staff's review of the SAR and other information submitted in support of the SAR, the NRC staff notes that DOE's descriptions of the programs for records, reports, tests, and inspections are reasonable and consistent with the guidance in the YMRP. The DOE program descriptions included (i) the program to be used to maintain the records; (ii) the program to report deficiencies to NRC; (iii) the program to perform, or permit NRC to perform, tests that NRC considers appropriate or necessary; and (iv) how DOE would address NRC inspections of the premises of the GROA and adjacent areas and provide access to DOE records, upon reasonable notice.

DOE Organizational Structure as it Pertains to Construction and Operation of Geologic Repository Operations Area

The SAR includes the organizational structure pertaining to construction and operation of the GROA at the Yucca Mountain site, and a description of any delegations of authority and assignments of responsibilities. DOE described the organizational structure anticipated at the time of repository construction and operations for the GROA at the Yucca Mountain site, including a description of a procedure for delegation of authority. DOE described the responsibilities of the director, management functions and responsibilities, reporting relationships, and principal lines of communication. DOE stated it would update the organizational structure to ensure that the nuclear criticality safety program is administratively independent of operations.

On the basis of the NRC staff's review of the SAR and other information submitted in support of the SAR, the NRC staff notes that DOE provided a reasonable description of the organizational structure for the construction and operation of the GROA that is consistent with the guidance in the YMRP.

Key Positions Assigned Responsibility for Safety and Operations of Geologic Repository Operations Area

The SAR includes the description for identification of the key positions that are assigned responsibility for safety at, and operation of, the GROA. DOE described key positions, the responsibilities and qualifications of those holding those positions, and how qualified alternates would be identified to act in the absence of DOE staff assigned to the key positions. DOE stated it would provide a revised description of the responsibilities for the radiation protection manager that incorporated the responsibilities and qualifications for the criticality safety manager.

On the basis of the NRC staff's review of the SAR and other information submitted in support of the SAR, the NRC staff notes that DOE provided a reasonable description of the key positions assigned responsibility for GROA safety and operations that is consistent with the guidance in the YMRP.

Personnel Qualifications and Training Requirements

The SAR includes a description of the personnel qualifications and training program. DOE described the management of the training function, identification of functional areas requiring training, objectives for training, organization training guides, and evaluation of trainee learning. DOE also described on-the-job training, personnel qualifications and certification, performance evaluations, physical condition of operational personnel, and quality assurance audits to determine training program effectiveness.

On the basis of the NRC staff's review of the SAR and other information submitted in support of the SAR, the NRC staff notes that DOE provided reasonable information on the personnel qualifications and training program that DOE stated will be implemented before DOE receives, possesses, stores, or disposes high-level radioactive waste. The DOE's information regarding the personnel qualifications and training program is consistent with the YMRP, which recognizes DOE is not expected to have an NRC-approved personnel qualifications and training program in place for a construction authorization.

Plans for Startup Activities and Testing

The SAR includes a description of plans for startup activities and startup testing in which DOE described

- The compatibility of testing programs with applicable regulatory guidance

- Use of experience from similar activities

- Test procedure development, approval by authorized personnel, and evaluation of test results

- Format and content of test procedures

- Component testing

- Systems functional testing

- Cold integrated systems testing

- Operational readiness review

- Protection of workers and the public

- Hot testing, which includes initial startup operations

- The schedules for startup activities and testing

- Testing and evaluating functional adequacy of new or untested systems, structures, and components

On the basis of the NRC staff's review of the SAR and other information submitted in support of the SAR, the NRC staff notes that DOE has reasonably described the plans for startup activities, which DOE stated will be implemented before DOE receives, possesses, stores, or disposes high-level radioactive waste. The DOE's description of its plans for startup activities and testing is consistent with the YMRP, which recognizes that DOE is not expected to have prepared plans for startup activities and testing for a construction authorization.

Plans for Conduct of Normal Activities, Including Maintenance, Surveillance, and Periodic Testing

The SAR includes a description of plans for conduct of normal activities, including maintenance, surveillance, and periodic testing of SSCs of the GROA. DOE described its plan and procedure development, testing, and approval by authorized personnel; management systems for operation of the repository, including administrative and procedural safety controls; and the specific types of plans and procedures to be developed for normal operations, maintenance, and periodic surveillance testing. DOE also identified experience from other, similar DOE facilities as guidance for developing plans and procedures for conduct of normal activities. DOE stated it would revise the description of the experience and competency for the independent reviewers of procedures.

On the basis of the NRC staff's review of the SAR and other information submitted in support of the SAR, the NRC staff notes that DOE reasonably described plans for conduct of normal activities, including maintenance, surveillance, and periodic testing that DOE stated will be implemented before DOE receives, possesses, stores, or disposes high-level radioactive waste. The DOE's description of its plans for normal activities, including maintenance, surveillance, and periodic testing, is consistent with the YMRP, which recognizes that DOE is not expected to have prepared plans in place for normal activities for a construction authorization

Emergency Planning

The SAR includes a description of the plan for responding to, and recovering from, radiological emergencies that may occur any time before permanent closure and decontamination or decontamination and dismantlement of surface facilities. DOE described its plan and stated it would provide its Emergency Plan to NRC no later than 6 months prior to the submittal of the updated application for a license to receive and possess high-level radioactive waste.

Although a detailed plan is not available at this time, on the basis of the NRC staff's review of the SAR and other information submitted in support of the SAR, the NRC staff notes that DOE's description of its plan for responding to, and recovering from, radiological emergencies that may occur any time before permanent closure and decontamination or decontamination and dismantlement of surface facilities is reasonable, in light of the information available.

Controls To Restrict Access and Regulate Land Uses

The SAR includes a description of the controls to restrict access and to regulate land uses at the Yucca Mountain site and adjacent areas, including a conceptual design of monuments that would be used to identify the site after permanent closure. DOE stated that (i) monuments and markers identify the GROA, postclosure controlled area, and preclosure controlled area; (ii) the conceptual design includes monument design, fabrication, and emplacement considerations to ensure monuments are as permanent as practicable; and (iii) the monuments will communicate information and warnings.

DOE described the steps under its control, taken to establish effective jurisdiction and control and acquisition or withdrawal of land area of the GROA. DOE identified the preclosure controlled area and described that access control and flight restrictions would be applied. DOE stated it would clarify landownership boundaries, the proposed land withdrawal boundary, and the boundary for the preclosure controlled area. DOE identified the postclosure controlled area and described its approach for postclosure controls. DOE described its approach to obtain water rights.

On the basis of the NRC staff's review of the SAR and other information submitted in support of the SAR, the NRC staff notes that DOE's description of the controls to restrict access and to regulate land uses is reasonable. The DOE's description of controls with respect to (i) additional controls for permanent closure, (ii) additional controls through permanent closure, and (iii) the conceptual design of monuments is consistent with the guidance in the YMRP. The NRC staff notes that DOE has taken the steps within its purview to obtain ownership of land and water rights; however, at this time, DOE does not have the ownership of the land or water rights.

Uses of the Geologic Repository Operations Area for Purposes Other Than Disposal of Radioactive Wastes

The SAR includes a description of the plans for any uses of the GROA for purposes other than radioactive waste disposal, with an analysis of the effects, if any, that such uses may have on the operation of the SSCs important to safety and the engineered and natural barriers important to waste isolation. DOE described potential other uses of the GROA and analyzed the effects that such uses may have on the operation of the SSCs important to safety and the engineered and natural barriers important to waste isolation. DOE described its procedures to manage the two ongoing other uses of the GROA (protection of cultural resources and protection of flora and fauna).

On the basis of the NRC staff's review of the SAR and other information submitted in support of the SAR, the NRC staff notes that DOE's description of the plans and procedures for activities other than waste disposal is reasonable. Additionally, DOE stated that its plans include an analysis of the effects, if any, that such uses may have on the operation of the SSCs important to safety and the engineered and natural barriers important to waste isolation. DOE's description of its plans for other uses is consistent with the guidance in the YMRP.

NRC Staff Conclusions

The U.S. Nuclear Regulatory Commission (NRC) staff reviewed the Safety Analysis Report and the other information submitted by the U.S. Department of Energy (DOE) and notes that the following information provided by DOE is reasonable:

- Description of the Research and Development Program

- Description of the Performance Confirmation Program

- Description of the Quality Assurance Program

- Descriptions of the programs for records, reports, tests, and inspections

- Description of the organizational structure

- Description of key positions

- Description of the personnel qualifications and training program

- Description of plans for startup testing

- Description of plans for conduct of normal activities

- Description of the plans for responding to and recovering from radiological emergencies

- Controls to restrict access and regulate land use (NRC staff notes that DOE has taken the steps within its purview to obtain ownership of land and water rights; however, at this time, DOE does not have the ownership of the land or water rights)

- Plans for uses of the geologic repository operations area (GROA) for purposes other than disposal of radioactive wastes

ACRONYMS AND ABBREVIATIONS

ANSI/ASME	American National Standards Institute/American Society of Mechanical Engineers
DOE	U.S. Department of Energy
EALs	emergency action levels
EBS	engineered barrier system
GROA	geologic repository operations area
HLW	high-level radioactive waste
HLWRS	High-Level Waste Repository Safety
ISG	interim staff guidance
NRC	U.S. Nuclear Regulatory Commission
OCRWM	Office of Civilian Radioactive Waste Management
PCSA	preclosure safety analysis
QARD	quality assurance requirements and description
RAI	request for additional information
RIPB	risk-informed, performance-based
SAR	Safety Analysis Report
SSCs	structures, systems, and components
TER	Technical Evaluation Report
TSPA	total system performance assessment
YMRP	Yucca Mountain Review Plan

INTRODUCTION

After docketing the U.S. Department of Energy (DOE) license application seeking a construction authorization for the proposed repository at Yucca Mountain, Nevada, the U.S. Nuclear Regulatory Commission (NRC) staff began documenting its review of the license application in a Safety Evaluation Report. On March 3, 2010, DOE filed a motion with the Atomic Safety and Licensing Board seeking to withdraw its license application to develop a repository at Yucca Mountain, Nevada. In June 2010, the Board denied the DOE motion. To date, petitions asking the Commission to reverse or uphold this decision are pending before the Commission.

On October 1, 2010, the NRC staff began commencing orderly closure of its Yucca Mountain activities. As part of orderly closure, the NRC staff prepared this technical evaluation report (TER), a knowledge management document. This document captures the NRC staff's technical assessment of information presented in DOE's Safety Analysis Report (SAR), dated June 3, 2008, as amended, and supporting information. The TER describes the NRC staff's technical evaluation of the DOE SAR, and, in particular, this Administrative and Programmatic Volume provides insights on administrative and programmatic activities used for ensuring safety in the context of geologic disposal. The TER was developed using the regulations at 10 CFR Part 63 and guidance in the Yucca Mountain Review Plan (YMRP). The TER does not, however, include conclusions as to whether or not DOE satisfies the Commission's regulations.

NRC's regulations at 10 CFR Part 63 provide site-specific criteria for geologic disposal at Yucca Mountain. These regulations prescribe requirements governing the licensing (including issuance of a construction authorization) of DOE to receive and possess source, special nuclear, and byproduct material at a geologic repository operations area sited, constructed, or operated at Yucca Mountain, Nevada. Under 10 CFR Part 63, there are several stages in the licensing process: the site characterization stage, the construction stage, and a period of operations. The period of operations includes the time during which emplacement would occur, any subsequent period before permanent closure during which the emplaced wastes are retrievable, and permanent closure. In addition, the regulations at 10 CFR Part 63 represent a risk-informed, performance-based (RIPB) approach to the review of geological disposal. The RIPB approach uses risk information to focus the review to areas most significant to safety or performance. The TER includes discussions regarding how the staff used risk information in its review. For the TER Administrative and Programmatic Volume, risk information was used only in the NRC's staff review of DOE's Performance Confirmation Program (Chapter 2 in this volume). In conducting its review, the NRC staff was guided by the Yucca Mountain Review Plan (YMRP).

This Administrative and Programmatic Volume evaluates DOE's research and development program for resolving safety questions that apply to systems, structures, and components (SSCs) important to safety and engineered and natural barriers important to waste isolation; DOE's Performance Confirmation Program; and administrative and programmatic management systems. The program for resolving safety questions identifies, describes, and discusses safety features or components that need further information to evaluate the design. DOE's Performance Confirmation Program examines the program of tests, experiments, and analyses DOE will conduct to evaluate the information used to support its postclosure performance assessments.

DOE addresses administrative and programmatic controls through management systems that control activities to ensure that the repository is designed, constructed, operated, and closed so

1

that high-level radioactive waste (HLW) or spent nuclear fuel is handled and emplaced such that the health and safety of workers and the public, and the environment will be protected. DOE's preclosure safety analysis (reviewed by NRC staff in the TER Preclosure Volume) determines which SSCs are important to safety. DOE stated that it will use management systems throughout the life of the repository to control activities and integrate programs to provide assurance that the repository will be constructed and operated within analyzed conditions and that the validity of the design and analytical bases is maintained as modifications occur. DOE's total system performance assessment (reviewed by NRC staff in the TER Postclosure Volume) provides an analytical basis for evaluating repository performance following closure. An analysis of the repository's natural and engineered barriers provides information on barrier features that are important to waste isolation. DOE stated that management systems ensure that (i) sufficient data exist to confirm its bases for the total system performance assessment are satisfied and (ii) the DOE Performance Confirmation Program provides confirmatory bases as part of making the determination to permanently close the repository. DOE stated that it will use procedural and administrative safety controls for important to safety and important to waste isolation SSCs to ensure they are maintained and operated within analyzed conditions, and are capable of performing their intended functions. DOE's management systems implement these administrative and procedural safety controls for activities affecting important to safety and important to waste isolation SSCs by providing the administrative and programmatic framework for

- Quality assurance
- Records, reports, tests, and inspection
- Personnel qualifications and training
- Startup activities and testing
- Conduct of normal activities, including maintenance, surveillance, and periodic testing
- Emergency planning
- Controls to restrict access and regulate land use
- Using the GROA for purposes other than disposal of radioactive waste

DOE stated that the management systems in each of these programs will be implemented through procedures governing work processes in accordance with its quality assurance program.

DOE's SAR provides the following information related to its administrative and programmatic program:

1. An identification of those structures, systems, and components of the geologic repository, both surface and subsurface, that require research and development to resolve safety questions

2. A description of the Performance Confirmation Program

3. A description of the Quality Assurance Program to be applied to the structures, systems, and components important to safety and to the engineered and natural barriers important to waste isolation

4. A description of the plan for responding to, and recovering from, radiological emergencies that may occur at any time before permanent closure and decontamination or decontamination and dismantlement of surface facilities

2

5.   The following information concerning activities at the GROA:

- The organizational structure of DOE as it pertains to construction and operation of the GROA, including a description of any delegations of authority and assignments of responsibilities, whether in the form of regulations, administrative directives, contract provisions, or otherwise

- Identification of key positions that are assigned responsibility for safety at and operation of the GROA

- Personnel qualifications and training program

- Plans for startup activities and startup testing

- Plans for conduct of normal activities, including maintenance, surveillance, and periodic testing of structures, systems, and components of the GROA

- Plans for any uses of the GROA at the Yucca Mountain site for purposes other than disposal of radioactive wastes, with an analysis of the effects, if any, that such uses may have on the operation of the SSCs important to safety and the engineered and natural barriers important to waste isolation

6.   A description of the program to be used to maintain records

7.   A description of the controls that DOE will apply to restrict access and to regulate land use at the Yucca Mountain site and adjacent areas, including a conceptual design of monuments that would be used to identify the site after permanent closure

The subsequent chapters in the TER Administrative and Programmatic Volume document the results of the NRC staff's technical evaluation of the SAR. The NRC staff also reviewed information DOE provided in response to NRC staff's requests for additional information. Chapter 1 evaluates DOE's description of a research and development program to resolve safety questions. Chapter 2 evaluates DOE's Performance Confirmation Program. Chapter 3 evaluates DOE's description of its quality assurance program. Chapter 4 evaluates DOE's description of the program to maintain records, report deficiencies, perform tests, and allow inspections. Chapter 5 evaluates DOE's organizational structure during construction. Chapter 6 evaluates DOE's plans for key positions for safety and operations at the geologic repository operations area. Chapter 7 evaluates DOE's description of its personnel qualifications and training program. Chapter 8 evaluates DOE's plans for startup activities and testing. Chapter 9 evaluates DOE's plans for conduct of normal activities. Chapter 10 evaluates DOE's description of the plans to respond to, and recovering from radiological emergencies that may occur at the geologic repository. Chapter 11 evaluates DOE's description of controls to restrict access and regulate land uses. Chapter 12 evaluates DOE's plans for uses of the geologic repository operations area for other purposes. Chapter 13 provides a summary of the conclusions of the NRC staff's technical evaluations of the administrative and programmatic aspects of the disposal facility. Chapter 14 provides a glossary of terms used in this volume of the TER.

CHAPTER 1

2.3 Research and Development Program To Resolve Safety Questions

2.3.1   Introduction

This chapter provides the U.S. Nuclear Regulatory Commission (NRC) staff's review of the U.S. Department of Energy's (DOE) research and development program to resolve safety questions. This review considers information provided in DOE's Safety Analysis Report (SAR) Chapter 3 (DOE, 2008ab, 2009av).

Safety questions related to either the design of structures, systems, and components (SSCs) of the geologic repository important to safety, both surface and subsurface, or related to natural and engineered barriers important to waste isolation may be identified by DOE in the SAR or by the NRC staff as a result of its SAR review. If safety questions are identified, DOE stated that it would implement a research and development program to resolve them. Such a program would be separate and distinct from DOE's Performance Confirmation Program described in SAR Chapter 4.

2.3.2   Evaluation Criteria

The requirement for research and development programs to resolve safety questions is specified in 10 CFR 63.21(c)(16). The SAR must include an identification of those SSCs of the geologic repository, both surface and subsurface, that require research and development to confirm the adequacy of design. For SSCs important to safety and for engineered and natural barriers important to waste isolation, DOE must provide a detailed description of the programs designed to resolve safety questions, including a schedule indicating when these questions would be resolved.

In its review, the NRC staff used the guidance in the Yucca Mountain Review Plan (YMRP) Section 2.3 (NRC, 2003aa). The YMRP acceptance criteria follow:

- DOE identification and descriptions of safety questions are adequate.

- DOE adequately identifies and describes in detail a research and development program that will be conducted to resolve safety questions, in a reasonable time period, for SSCs important to safety and engineered and natural barriers important to waste isolation.

- DOE provides a reasonable schedule for the completion of the program relative to the projected startup date of repository operations and makes a commitment to include resolved safety questions in requested amendments to the license application, as appropriate.

- DOE provides the design alternatives or operational restrictions available if the results of the program do not demonstrate acceptable resolution of the problem.

2.3.3 Technical Evaluation

2.3.3.1 Identification and Description of Safety Questions

DOE did not identify any safety questions with respect to (i) SSCs important to safety or (ii) engineered and natural barriers important to waste isolation (SAR Section 3.1).

The NRC staff review of the DOE SAR did not identify any safety questions with respect to (i) SSCs important to safety or (ii) engineered and natural barriers important to waste isolation.

2.3.3.2 Research and Development Programs Related to Safety Questions

A specific research and development program would be developed if safety issues were identified and would be separate and distinct from DOE's Performance Confirmation Program described in SAR Chapter 4. Results of a research and development program(s), including periodic progress updates, would be provided to NRC (SAR Chapter 3.2). Should safety questions be identified in the future, DOE stated that the program would

- Identify and describe safety questions

- Identify and describe the research and development that will be conducted to resolve safety questions

- Provide a schedule for completing the activities relative to the projected startup date of repository operations

- Provide design alternatives or operational restrictions available if the results of program activities do not resolve the safety questions

2.3.4 NRC Staff Conclusions

The NRC staff has reviewed the SAR and the other information submitted in support of the SAR. Because neither DOE in its SAR nor the NRC staff in its review of the SAR identified any safety questions, the NRC staff notes that a research and development program that is separate and distinct from DOE's Performance Confirmation Program is not needed at this time. The NRC staff notes that DOE's description of its approach for developing and implementing a research and development program is consistent with the guidance in the YMRP and is reasonable as a general approach for use in addressing safety questions should a safety question be identified in the future.

2.3.5 References

DOE. 2009av. DOE/RW–0573, "Yucca Mountain Repository License Application." Rev. 1. ML090700817. Las Vegas, Nevada: DOE, Office of Civilian Radioactive Waste Management.

DOE. 2008ab. DOE/RW–0573, "Yucca Mountain Repository License Application." Rev. 0. ML081560400. Las Vegas, Nevada: DOE, Office of Civilian Radioactive Waste Management.

NRC. 2003aa. NUREG–1804, "Yucca Mountain Review Plan—Final Report." Rev. 2. ML032030389. Washington, DC: NRC.

CHAPTER 2

2.4 Performance Confirmation Program

2.4.1 Introduction

This chapter evaluates the description of the Performance Confirmation Program provided in the U.S. Department of Energy (DOE) Safety Analysis Report (SAR) Chapter 4 (DOE, 2008ab, 2009av). DOE provided more information on this topic in its response to NRC staff requests for additional information (DOE, 2009an,gm, 2010ap).

The purpose of the Performance Confirmation Program is to evaluate and confirm the assumptions, data, and analyses that support the performance of the repository. Key geotechnical and design parameters, including any interactions between natural and engineered systems and components, are to be monitored by DOE throughout site characterization, construction, emplacement, and operation to identify any significant changes in the conditions assumed in the SAR that may affect postclosure safety.

The focus of the Performance Confirmation Program is on subsurface conditions, as well as the natural and engineered systems and components that DOE designed or assumed to operate as barriers after permanent closure. The Performance Confirmation Program does not include testing and monitoring to confirm preclosure performance in other contexts (i.e., testing and monitoring structures, systems, and components important to safety); however, the Performance Confirmation Program should consider retrievability when monitoring subsurface conditions.

2.4.2 Evaluation Criteria

10 CFR 63.21(c)(17) requires that the SAR include a description of the Performance Confirmation Program that meets the requirements of 10 CFR Part 63, Subpart F. 10 CFR 63.74 provides the requirements for tests, and 10 CFR 63.74(b) specifies that these tests are to include a Performance Confirmation Program carried out in accordance with 10 CFR Part 63, Subpart F.

10 CFR Part 63, Subpart F specifies requirements in four areas: (i) general requirements (10 CFR 63.131), (ii) confirmation of geotechnical and design parameters (10 CFR 63.132), (iii) design testing (10 CFR 63.133), and (iv) monitoring and testing waste packages (10 CFR 63.134).

The general requirements are specified in 10 CFR 63.131(a)–(d). 10 CFR 63.131(a) requires that the Performance Confirmation Program must provide data that indicate, where practicable, whether (i) actual subsurface conditions encountered and changes in those conditions during construction and waste emplacement operations are within limits assumed in the review and (ii) natural and engineered systems and components required for repository operation, and that are designed or assumed to operate as barriers after permanent closure, are functioning as intended and anticipated. 10 CFR 63.131(b) requires that the program start during site characterization and continue until permanent closure. 10 CFR 63.131(c) requires that the program include *in-situ* monitoring, laboratory and field testing, and *in-situ* experiments, as may be appropriate to provide the data required by 10 CFR 63.131(a). 10 CFR 63.131(d) requires that the program be implemented so that (i) it does not adversely affect the geologic and engineered elements of the geologic repository; (ii) it provides baseline information and analysis of that information on those parameters and natural processes pertaining to the geologic setting that may be changed by site characterization, construction, and operational activities; and (iii) it monitors and analyzes changes from the baseline condition of parameters that could affect the performance of a geologic repository. 10 CFR Part 63 specifies requirements for confirmation of geotechnical and design parameters in 10 CFR 63.132(a)–(e), design testing in 10 CFR 63.133(a)–(d), and monitoring and testing waste packages in 10 CFR 63.134(a)–(d).

The NRC staff evaluated DOE's description of its Performance Confirmation Program, using guidance in YMRP Section 2.4. The acceptance criteria follow:

- The Performance Confirmation Program meets the general requirements established for such a program.

- The Performance Confirmation Program to confirm geotechnical and design parameters meets the requirements established for such a program.

- The Performance Confirmation Program for design testing meets the requirements established for such a program.

- The Performance Confirmation Program for monitoring and testing waste packages meets the requirements established for such a program.

2.4.3   Technical Evaluation

The NRC staff reviewed DOE's description of the Performance Confirmation Program and, as necessary, additional information describing the Performance Confirmation Program in the Performance Confirmation Plan (SNL, 2008aq) using the guidance in the YMRP.

2.4.3.1   Performance Confirmation Program Planning

The NRC staff's technical evaluation of the Performance Confirmation Program planning is discussed in the following five sections: (i) objectives of the Performance Confirmation Program (TER Section 2.4.3.1.1), (ii) schedule for the Performance Confirmation Program (TER Section 2.4.3.1.2), (iii) implementation of the Performance Confirmation Program (TER Section 2.4.3.1.3), (iv) records and reports related to the Performance Confirmation Program, and (v) NRC staff evaluation of the Performance Confirmation Program planning (TER Section 2.4.3.1.5).

2.4.3.1.1   Objectives of the Performance Confirmation Program

DOE described the objectives of the Performance Confirmation Program in SAR Section 4.1. Specifically, DOE stated the program is designed to confirm the adequacy of assumptions, data, and analyses that support the postclosure safety determinations. DOE stated that the Performance Confirmation Program evaluates the information supporting the performance assessments for individual protection and groundwater protection, as well as consideration of preclosure aspects of repository performance, such as retrievability (SAR p. 4-3).

DOE identified two specific objectives for the Performance Confirmation Program. First, the program provides information, where practicable, to confirm subsurface conditions encountered during construction and waste emplacement operations and changes in those conditions are within the conditions assumed in the SAR. This includes monitoring subsurface conditions and tests to confirm geotechnical and design assumptions for retrievability (NRC staff's review of retrievability is documented in TER Section 2.1.2). Second, the program provides information to confirm that the natural and engineered barriers are functioning as described in SAR Chapter 2.

Identification of Natural and Engineered Systems and Components Operating as Barriers

The NRC staff review first considered whether DOE's identification of barriers with respect to the Performance Confirmation Program is reasonable. Second, NRC staff considered whether DOE identified (i) the natural and engineered systems and components to be monitored and tested to ensure that they are functioning as intended and (ii) the specific geotechnical and design parameters to be measured or observed. DOE identified the natural and engineered barriers that are important to waste isolation (i.e., those that prevent or substantially reduce the rate of movement of water or radionuclides from the repository to the accessible environment or that prevent the release or substantially reduce the release rate of radionuclides from the waste) in SAR Table 4-1 and related them to the performance confirmation activities. In addition, DOE provided more detail with respect to the relevant (i) barrier; (ii) feature, event, or process; (iii) effect on barrier capability; and (iv) core parameter characteristic for each performance confirmation activity in its Performance Confirmation Plan, addendum to Revision 5, Table A–2[a]. DOE's description of barriers, as related to the Performance Confirmation Program, is reasonable because the description is consistent with the identification and description of barriers and their capabilities presented in SAR Section 2.1, including SAR Table 2.1-1 (expanded; DOE, 2009an), which was reviewed in the TER Postclosure Volume, Section 2.2.1.1.

Identification of Geotechnical and Design Parameters To Be Measured or Observed

As described in SAR Section 4.1, the Performance Confirmation Plan identifies 20 activities for performance confirmation. In SAR Table 4-1, DOE provided for each type of monitoring or testing; the candidate activities, including a description and purpose; the candidate parameters for each candidate activity; and the related barrier or event. For example, for the seepage monitoring candidate activity, DOE identified that (i) the activity description is seepage monitoring and laboratory analysis of water samples; (ii) the candidate parameters include seepage rate, locations, and quantity and chemical composition; (iii) the purpose is to evaluate results from the seepage model; and (iv) the related barrier is the upper natural barrier.

In SAR Table 4-2, DOE identified the eight performance confirmation activities to confirm subsurface conditions and changes in subsurface conditions during construction and waste emplacement operations. Also in SAR Table 4-2, DOE identified the 19 performance confirmation activities to confirm that the natural and engineered barriers are functioning as described in SAR Chapter 2. DOE identified, in SAR Tables 4-1 and 4-2, the natural and engineered systems and components selected to be monitored and tested to ensure that they are functioning as intended and identified the specific geotechnical and design parameters selected to be measured or observed.

In SAR Table 4-2, DOE identified that each performance confirmation activity may accomplish more than one objective. The NRC staff's evaluation of the individual performance confirmation activities that were identified as related to geotechnical and design parameters, design testing, and monitoring and testing of waste packages is in TER Sections 2.4.3.2, 2.4.3.3, and 2.4.3.4, respectively. Those individual performance confirmation activities identified as related to precipitation monitoring, subsurface water and rock testing, unsaturated zone testing, saturated zone monitoring, saturated zone fault hydrology testing, and saturated zone alluvium testing are reviewed in TER Sections 2.4.3.1.1.1–2.4.3.1.1.6, respectively.

<u>Methodology Used To Select the Natural and Engineered Systems Components, and the Geotechnical and Design Components</u>

In SAR Section 4.1.1, DOE described eight principles (e.g., one principle is to confirm the bases relied upon for retrieval of waste) that it applies to the administration of its Performance Confirmation Program and summarized its methodology for selecting performance confirmation activities. DOE stated that the performance confirmation activities were selected using a risk-informed, performance-based methodology and that the details on the decision analysis are in the Performance Confirmation Plan Section 1.4.1. DOE stated that it used a methodological approach (multi-attribute utility analysis) to identify relevant geotechnical and design parameters and determine appropriate testing activities. DOE's decision analysis was based on its understanding of the performance assessment and barrier capability existing prior to completing the total system performance assessment (TSPA) (SNL, 2008ag) presented in the SAR. DOE's approach consisted of (i) using subject matter experts to identify relevant geotechnical and design parameters and (ii) determining appropriate testing activities on the basis of the application of the following three criteria:

- Sensitivity of barrier capability and system performance to the parameter
- Level of confidence in the current knowledge about the parameter
- Accuracy of information obtained by a particular test

DOE stated the performance confirmation emphasized parameters related to barriers important to waste isolation. Because the models used to develop the TSPA have been updated since the multi-attribute utility analysis was used to identify the performance confirmation activities, DOE conducted an additional completeness review comparing the planned activities to the representation of key features and processes in the final TSPA models as described in the Performance Confirmation Plan addendum, Appendix A[a]. DOE compared the final TSPA presented in the SAR, the postclosure nuclear safety design bases (SNL, 2008ad), and the Performance Confirmation Plan. DOE did not identify new performance confirmation activities as a result of its additional completeness review.

DOE's method to select geotechnical and design parameters to measure or observe and DOE's method to select the natural and engineered systems and components to monitor and test are reasonable because (i) the methods are risk informed and performance based; (ii) DOE assessed its selection of performance confirmation activities, derived from the multi-attribute analysis, against the final TSPA in the SAR to determine whether any changes in activities were needed; and (iii) the methodology is consistent with the DOE's technical bases in the SAR, including the natural and engineered barrier system (EBS) and components.

<u>*In-Situ* Monitoring, Laboratory and Field Testing, and *In-Situ* Experiments</u>

DOE provided information in SAR Table 4-1 on the *in-situ* monitoring, laboratory and field testing, and *in-situ* experiments it will use to acquire needed data and that it will apply to the selected geotechnical and design parameters and to the selected natural and engineered systems and components. In SAR Section 4.2, DOE provided the purpose, description of current understanding, and methodology for each of the candidate performance confirmation activities. For example, for the seepage monitoring candidate activity, DOE specified that seepage monitoring and sampling will include *in-situ* monitoring occurrences and quantities, as well as conducting laboratory analyses of seepage fluids. DOE described that its monitoring will be conducted at appropriate locations in the subsurface, and specific tests (field tests) will be conducted in unventilated alcoves or boreholes and in a thermally accelerated test drift (an *in-situ* experiment). The Performance Confirmation Plan provides a more in-depth discussion of these performance confirmation activities, including the purpose, description of current understanding, and candidate geotechnical and design parameters, including planned testing and monitoring methods and techniques.

Expected Changes from Baseline

In SAR Section 4.1.3, DOE addressed evaluation of results and reporting, and described how design basis information and baseline values are used in the Performance Confirmation Program. DOE stated that for geotechnical and design parameters, the baseline values will be derived from reference design basis documentation. For initial performance confirmation evaluations, the baseline data will be derived from analysis and model reports and performance assessment input parameters; baseline values will be used to define objectives for performance confirmation investigations, which are described in appropriate performance confirmation test plans. DOE described that sources for baseline values for selected parameters, test completion criteria, and variance criteria will be identified in performance confirmation test plans. Finally, DOE described how it intends to address performance confirmation results that exceed condition limits established in the performance confirmation test plans.

DOE's Performance Confirmation Program reasonably includes the expected changes (i.e., design bases and assumptions) from baseline for the selected geotechnical and design parameters, including natural processes, pertaining to natural systems and components that are assumed to operate as barriers after permanent closure because (i) DOE's performance confirmation test plans (SNL, 2007bo,bp; BSC, 2006al) described how design basis information and baseline values are used and identified the condition limits used in the Performance Confirmation Program and (ii) DOE described how it will address performance confirmation results that exceed condition limits.

Summary of NRC Staff Evaluation of Performance Confirmation Plan Objectives

The NRC staff reviewed DOE's description of the Performance Confirmation Program in SAR Section 4.2 and the Performance Confirmation Plan and notes that the Performance Confirmation Plan provides reasonable technical information and plans for *in-situ* monitoring, laboratory and field testing, and *in-situ* experiments to carry out the objectives because (i) DOE identified the natural and engineered systems and components that are designed or assumed to operate as barriers after permanent closure, including their specific functions, that it selected to monitor and test to ensure they are functioning as intended and expected; (ii) DOE identified specific parameters pertaining to natural systems and components that are assumed to operate as barriers after permanent closure, including natural processes, and any interactions between natural and engineered systems and components that it selected to be measured or observed; (iii) DOE included specific *in-situ* monitoring, laboratory and field testing, and *in-situ* experiments to acquire needed data; (iv) DOE specified which *in-situ* monitoring, laboratory and field testing, or *in-situ* experimental methods it will apply to the selected parameters pertaining to natural and engineered systems and components that are designed or assumed to operate as barriers after permanent closure and interactions between natural and engineered systems and components; and (v) DOE included the expected changes (i.e., design bases and assumptions) from the baseline for the selected geotechnical and design parameters, and natural processes.

In TER Sections 2.4.3.1.1.1–2.4.3.1.1.6, the NRC staff reviews the individual performance confirmation activities that DOE identified for precipitation monitoring, subsurface water and rock testing, unsaturated zone testing, saturated zone monitoring, saturated zone fault hydrology testing, and saturated zone alluvium testing activities focused on the selected parameters (either candidate or final). The DOE phased approach (i.e., final parameters are identified when performance confirmation test plans are completed) provides the NRC staff the opportunity to review and approve final parameters prior to implementation of the test plans (see TER Section 2.4.3.1.3 regarding implementation).

2.4.3.1.1.1   Precipitation Monitoring

DOE described precipitation monitoring in SAR Section 4.2.1.1, DOE (2009gm), Performance Confirmation Plan Section 3.3.1.1, and SNL (2007bp). DOE stated the purpose of precipitation monitoring is to confirm the information used in conceptual and numerical models of the hydrologic conditions described in SAR Section 2.3.1. DOE also stated (i) precipitation represents the predominant input of water into the upper natural barrier, (ii) the information collected for this activity will confirm and extend the precipitation record for the site and be used for comparison with seepage observations, and (iii) the performance of the upper natural barrier may be supported by comparing the precipitation inputs with observed fluxes at depth. Precipitation monitoring began during site characterization and will continue through closure using six existing monitoring stations to support development of the infiltration model. The methodology includes instrumentation of the six sites with two precipitation gauges: one for precipitation rate and one for precipitation quantity.

Although DOE identified precipitation rate, quantity, and chemical composition as candidate parameters in SAR Table 4-1 and Performance Confirmation Plan Table 3-2, DOE (2009gm) clarified that test plans specify parameters that will meet the objectives of the activity and that the precipitation monitoring test plan (SNL, 2007bp) is complete; thus, DOE's monitoring parameters are precipitation rate and precipitation quantity. The NRC staff notes that these parameters are reasonable because they are designed to provide similar types of data used in the process model, from the same or similar locations used during site characterization, to confirm the precipitation input data used in the infiltration model.

DOE's methodology to establish a baseline for the parameters is reasonable because baseline data were derived from the analysis supporting the infiltration model and are included in the detailed performance confirmation test plan. The baseline (i.e., expected range) for parameters in SNL Table 1-2 (2007bp) is reasonable because it is consistent with information used to develop the infiltration analysis in the SAR (SNL, 2007az). The NRC staff notes that the methodology is reasonable for the parameters because it is consistent with the techniques used since site characterization and will collect applicable information to confirm DOE assumptions and parameters used in the TSPA.

2.4.3.1.1.2   Subsurface Water and Rock Testing

The subsurface water and rock testing activity is described in SAR Section 4.2.1.3 and in Performance Confirmation Plan Section 3.3.1.3. The purpose of this activity is to evaluate whether the upper natural barrier operates as expected and to confirm actual subsurface conditions encountered, thereby verifying assumptions for fast flow paths used in unsaturated zone models. DOE plans to analyze pore-water samples obtained from the rock cores for dissolved ions, analyze the core for uranium and strontium isotopes, and obtain fracture coatings from within the drifts and analyze the coatings for isotope geochemistry. DOE initiated sampling and laboratory analysis of water, rock, and fracture-filling materials during site characterization, using the data to infer present and historical percolation fluxes at selected locations (SAR Section 2.3.2), and stated it will continue this practice throughout repository construction. In SAR Table 4-1 and Performance Confirmation Plan Table 3-2, DOE identified that chloride concentration; isotopic composition for U, Sr, and O; H-3; Cl/Cl-36; Tc-99; and I-129/I-127 are the candidate parameters.

DOE's Performance Confirmation Plan for subsurface water and rock testing is reasonable because (i) the candidate parameters DOE selected for the testing program are designed to confirm information used by DOE to describe percolation fluxes in the unsaturated zone, (ii) baseline data will be derived from existing performance assessment input data and information in analysis and model reports and will be included in the detailed performance confirmation test plan, and (iii) the testing program utilizes available techniques that have been used since site characterization.

2.4.3.1.1.3   Unsaturated Zone Testing

DOE described the unsaturated zone testing activity in SAR Section 4.2.1.4 and Performance Confirmation Plan Section 3.3.1.4. The purpose of this activity is to confirm that sorption coefficients in the rock below the repository are within established limits used in performance assessment models. DOE plans to use unsaturated zone testing to evaluate transport properties and field sorptive properties of the Topopah Spring Tuff crystal-poor member in ambient seepage alcoves or drifts with no waste packages emplaced. This activity includes *in-situ* experiments, field mapping, field testing, and laboratory analysis of samples collected from the field tests. DOE identified that transport and sorption testing will be conducted in two or more seepage monitoring alcoves located within the repository. Similar activities were performed during site characterization to characterize comparable parameters in nonwelded tuffs below the repository. DOE stated that unsaturated zone testing will begin during construction and continue to the early stages of the emplacement period. In SAR Table 4-1 and Performance Confirmation Plan Table 3-2, DOE identified sorption parameters, van Genuchten parameters describing fractures and matrix, colloid/colloid-facilitated transport parameters, fracture density, apertures, coatings, air permeability, seepage, alcove temperature, and relative humidity as the candidate parameters.

DOE's Performance Confirmation Plan for the unsaturated zone testing is reasonable because (i) the candidate parameters DOE selected for the testing program are designed to confirm information used by DOE to describe transport and sorption in the unsaturated zone models supporting the performance assessment, (ii) DOE plans to synthesize existing baseline data from performance assessment results and information in analysis and model reports and baseline data will be presented in the detailed performance confirmation test plan, and (iii) the testing program utilizes available techniques that have been used since site characterization.

2.4.3.1.1.4   Saturated Zone Monitoring

DOE described the saturated zone monitoring activity in SAR Section 4.2.1.5 and Performance Confirmation Plan Section 3.3.1.5. The purpose of this activity is to evaluate hydrologic and chemical parameters used with the saturated zone flow model and includes monitoring the absence of repository radionuclides in downgradient wells and arrival of radionuclides from upgradient sources, such as nuclear testing. Saturated zone monitoring is described as measuring water levels, Eh, and pH in site and Nye County wells, and analyzing radionuclide concentrations in water samples obtained from the wells. This activity began during site characterization, and DOE stated that it will continue until permanent closure. In SAR Table 4-1 and Performance Confirmation Plan Table 3-2, DOE identified water level and hydrochemical indicators (Eh, pH, radionuclide concentrations, and colloid characteristics) as the candidate parameters.

DOE's Performance Confirmation Plan for the saturated zone testing is reasonable because (i) the candidate parameters DOE selected for the testing program are designed to provide information indicating whether the EBS and lower natural barrier are functioning as anticipated, (ii) DOE plans to synthesize baseline data from performance assessment results and information in analysis and model reports and baseline data will be included in the detailed performance confirmation test plan, and (iii) the testing program utilizes available techniques that have been used since site characterization.

2.4.3.1.1.5   Saturated Zone Fault Hydrology Testing

DOE described saturated zone fault hydrology testing activity in SAR Section 4.2.1.6 and in Performance Confirmation Plan Section 3.3.1.6. The purpose of this activity is to evaluate fault parameter assumptions used in the saturated zone flow and transport models. DOE described the tuff portion of the saturated zone barrier as complicated by faulting and tilting, with faults acting as both barriers to and preferential pathways for flow. The planned tests are described as similar to tests previously performed at the C-well testing complex, perhaps including monitoring of water levels, single borehole and cross-hole hydraulic and tracer tests, field sample collection, and laboratory analysis of samples. DOE stated that it plans to drill additional boreholes in or near faults to perform the testing and identified the Solitario Canyon Fault system and an undetermined location downgradient of the repository as candidate test locations, with candidate formations including the Tertiary tuff, the Crater Flat Group, and the Paintbrush Group. DOE described this new activity as relatively short term, with several phases each between 1 and 3 years in duration, and that it will be initiated during construction. In SAR Table 4-1 and in Performance Confirmation Plan Table 3-2, DOE identified transmissivity; hydraulic conductivity; water flux and specific discharge; effective flow porosity; longitudinal dispersivity; sorption parameters; parameters describing diffusion between flowing and stagnant water; colloid or colloid-facilitated transport parameters; Eh; pH; and natural colloid concentrations, including anisotropy, as the candidate parameters.

DOE's Performance Confirmation Plan for the saturated zone fault hydrology testing is reasonable because (i) the candidate parameters DOE selected for the testing program are similar to the set of parameters DOE used to characterize the hydraulic and transport characteristics of tuff aquifers in previous testing at the C-well testing complex during site characterization, (ii) DOE plans to synthesize baseline data from performance assessment results and published results from analog sites in fractured and faulted rocks and the baseline data will be included in the detailed performance confirmation test plan, and (iii) the testing program utilizes available techniques that are consistent with techniques used since site characterization and will collect applicable information to confirm DOE assumptions and parameters used in the TSPA.

2.4.3.1.1.6   Saturated Zone Alluvium Testing

DOE described the saturated zone alluvium testing activity in SAR Section 4.2.1.7, Performance Confirmation Plan Section 3.3.1.7, and DOE (2009gm). The purpose of this activity is to confirm inputs and assumptions for the saturated zone and flow models. The activity includes testing and monitoring of the alluvium to evaluate the assumptions and results of conceptual and numerical models describing saturated zone hydrologic conditions in the alluvium south of the site. DOE plans to perform testing at the existing Alluvial Testing Complex, approximately located at the boundary of the accessible environment, which is also the site of hydraulic testing performed during site characterization. DOE stated its planned tests will include monitoring of water levels, single borehole and cross-hole hydraulic and tracer tests in the saturated portion of the alluvium, field sample collection, and laboratory analysis. DOE also plans to conduct laboratory batch and column sorption tests to compare the sorption properties of tracers and radionuclides. Similar testing was performed during site characterization. As DOE (2009gm) described, testing associated with this activity is expected to be between 1 and 3 years in duration and could be resumed at any time; however, DOE stated that the performance confirmation test plan will be completed during construction. In SAR Table 4-1 and Performance Confirmation Plan Table 3-2, DOE identified transmissivity, hydraulic conductivity, water flux and specific discharge, effective flow porosity, longitudinal dispersivity, sorption parameters, parameters describing diffusion between flowing and stagnant water, colloid or colloid-facilitated transport parameters, Eh, pH, and natural colloid concentrations as the candidate parameters.

DOE's Performance Confirmation Plan for the saturated zone alluvium testing is reasonable because (i) the candidate parameters DOE selected for the testing program are similar to the set of parameters DOE used in evaluating hydraulic and transport characteristics of the alluvial aquifer in previous testing at the Alluvial Testing Complex during site characterization, (ii) DOE plans to synthesize baseline data from existing performance assessment results and information in analysis and model reports and the baseline data will be included in the detailed performance confirmation test plan, and (iii) the testing program utilizes available techniques consistent with the techniques used since site characterization and will collect applicable information to confirm DOE assumptions and parameters used in the TSPA.

2.4.3.1.2   Schedule for the Performance Confirmation Program

DOE described the duration of the Performance Confirmation Program in SAR Section 4.1 and provided a schedule for planned activities in SAR Figure 4-2. DOE stated that the program began during site characterization, assumes a 100-year preclosure period for performance confirmation activities, and will continue until permanent closure. The NRC staff notes that DOE's schedule for its planned performance confirmation activities is reasonable because it provides for testing that will continue until permanent closure.

2.4.3.1.3   Implementation of the Performance Confirmation Program

DOE provided a program overview in SAR Section 4.1; addressed implementation of the Performance Confirmation Program in SAR Sections 4.1.1 and 4.1.2; and provided detailed information related to data management, analysis, and reporting, as well as test planning and implementation, in its Performance Confirmation Plan Sections 4 and 5. SAR Figure 4-1 presented the planning and procedural documents' hierarchy relevant to implementation of performance confirmation monitoring and testing. DOE has a phased approach for implementing its Performance Confirmation Program.

The Performance Confirmation Plan identifies 20 activities for performance confirmation. In SAR Section 4.1, DOE stated that the current conceptual descriptions of these activities in the Performance Confirmation Plan will be supplemented by performance confirmation test plans that provide the rigor necessary to justify the activity, plan the details of its implementation, and establish condition limits for results that indicate significant differences from baseline information for those geotechnical and design parameters to be evaluated through observation and measurement during construction and operation (SAR p. 4-3). Also, DOE identified that performance confirmation test plans have been written for seismic monitoring (SNL, 2007bo), precipitation monitoring (SNL, 2007bp), and construction effects monitoring (BSC, 2006al); other test plans will be prepared sequentially; and the Performance Confirmation Plan will be revised and updated as program development continues. DOE (2009gm) stated that future performance confirmation test plans will be provided to NRC at first issuance, prior to test implementation. In its Performance Confirmation Plan, DOE stated the Performance Confirmation Program must be flexible, with specific details of the program evolving as necessary in response to information obtained from performance confirmation activities.

As stated in TER Section 2.4.3.1.1, the NRC staff notes that DOE reasonably specified in SAR Table 4-1 the important geotechnical and design parameters to be evaluated through observation and measurement during construction and operation. In SAR Table 4-1 the parameters are identified as candidate parameters. DOE stated that candidate parameters remain preliminary until they are finalized in the performance confirmation test plans and that DOE will provide future performance confirmation test plans to NRC when the plans are completed. The DOE phased approach regarding identification of the final parameters provides the NRC staff the opportunity to review and approve those parameters prior to DOE's implementation. The NRC staff's review of the parameters identified in the precipitation monitoring test plan is provided in TER Section 2.4.3.1.1.1, and the seismic monitoring and construction effects monitoring test plans are reviewed in TER Section 2.4.3.2.

Procedures for Consideration of Adverse Effects

DOE provided information [SAR Section 4.1.2; Performance Confirmation Plan Sections 5.2.2 and 5.2.3; DOE (2009gm)] related to the procedures to consider adverse effects on the natural and engineered elements of the geologic repository before initiating any *in-situ* monitoring, test, or experiment to acquire data. The NRC staff notes that DOE has procedures to consider adverse effects because DOE (2009gm) stated that test construction and performance confirmation activities will be evaluated for their impact to waste isolation prior to test implementation and these evaluations will be documented separately for performance confirmation test plans under the Site Performance Protection Evaluation Program.

Providing Baseline Information and Monitoring and Analyzing Changes from the Baseline

In SAR Section 4.1.1, DOE stated that the Performance Confirmation Program is designed to (i) provide baseline information and analysis of that information on parameters and natural processes pertaining to the geologic setting that may be changed by site characterization, construction, or operational activities; (ii) monitor and analyze changes from baseline conditions that could affect repository design or performance; and (iii) monitor engineered systems and components intended to operate as barriers after permanent closure to ensure that they function, as assumed in the performance assessment. As described in SAR Section 4.1.2, the Performance Confirmation Plan contains general reporting requirements for testing and monitoring results (summarized in SAR Section 4.1.3), identification of testing activities (summarized in SAR Section 4.2), and testing and proposed monitoring methodology and techniques (summarized in SAR Section 4.2). DOE stated in SAR Section 4.1.2 that the performance confirmation test plans will provide detailed information on 20 items, including (i) baseline information; (ii) anticipated changes to be observed or measured during the period of the tests, including those that may be changed by site investigations, construction, and operations; and (iii) identification of what constitutes trends or variations beyond the anticipated range during the monitoring or testing period. DOE did not identify any exceptions to its plans to (i) monitor and analyze changes from baseline conditions that could affect repository design or performance and (ii) monitor engineered systems and components intended to operate as barriers after permanent closure to ensure that they function, as assumed in the performance assessment.

Summary of NRC Staff Evaluation of Implementation of Performance Confirmation Program

The NRC staff reviewed information in SAR Sections 4.1.1 and 4.1.2, the Performance Confirmation Plan, and the available performance confirmation test plans (SNL, 2007bo,bp; BSC, 2006al). DOE's implementation of its Performance Confirmation Program is reasonable because (i) DOE reasonably specified in SAR Table 4-1 the important geotechnical and design parameters to be evaluated through observation and measurement during construction and operation; (ii) the performance confirmation test plans will provide detailed information on the final parameters in the test plans; (iii) DOE stated that candidate parameters remain preliminary until they are finalized in the performance confirmation test plans and will provide future performance confirmation test plans to NRC when the plans are completed; (iv) DOE has procedures to consider adverse effects; (v) DOE includes plans for providing baseline information and analysis and monitoring and analyzing changes from baseline conditions that could affect repository design or performance; (vi) DOE includes plans for monitoring engineered systems and components intended to operate as barriers after permanent closure to ensure that they function, as assumed in the performance assessment; and (vii) DOE will evaluate the impact of test construction and performance confirmation activities on waste isolation prior to test implementation and these evaluations will be documented separately for performance confirmation test plans under the Site Performance Protection Evaluation Program. Additionally, as part of implementation, DOE stated that it plans to update periodically the Performance Confirmation Plan to ensure the information therein is consistent with the SAR and reflects the most current understanding of the TSPA.

2.4.3.1.4   Records and Reports Related to the Performance Confirmation Program

In SAR Section 4.1.2, DOE described program documentation and the hierarchical document structure that supports the Performance Confirmation Program and guides its execution. DOE stated that the scope and implementation of the Performance Confirmation Plan will be periodically assessed to evaluate its continued relevance. In SAR Section 4.1.2, DOE stated that applicable procedures common to program administration are summarized in the Performance Confirmation Plan, with details provided in performance confirmation test plans and supporting documents. Also, DOE described that the Performance Confirmation Plan, performance confirmation test plans, and detailed implementation documents will provide information for general reporting requirements for testing and monitoring results. DOE stated that records generated as a result of performance confirmation test implementation will be handled, stored, submitted, and retained in accordance with applicable existing procedures, and data resulting from performance confirmation testing and monitoring activities will be managed in accordance with applicable records procedures. Specific procedures are cited in the Performance Confirmation Plan.

In SAR Section 4.1.3, DOE described its evaluation of results and reporting. DOE stated that performance confirmation results that exceed condition limits are the basis for initiating procedures that lead to NRC notification and described the evaluation report that would be subsequently submitted. Also DOE stated in SAR Section 4.1.3 that recommended changes in construction design or performance assessment approaches will be reported to NRC.

DOE has reasonably described its approach for documenting and maintaining records and reports related to the Performance Confirmation Program because DOE (i) described how records will be handled, stored, submitted, and retained; (ii) described the information that would be included in reports to the NRC including recommended changes in construction design or performance assessment approaches; and (iii) provided the procedures for reports in the Performance Confirmation Plan.

2.4.3.1.5   Summary of the NRC Staff Evaluation on Performance Confirmation Program Planning

On the basis of the NRC staff's evaluation in TER Sections 2.4.3.1.1–2.4.3.1.4, the NRC staff notes that the information DOE provided on the performance confirmation planning is reasonable because of the following:

- The objectives of the Performance Confirmation Program will provide data to indicate whether (i) actual subsurface conditions encountered and changes in those conditions during construction and waste emplacement operations are within the limits assumed in the SAR and (ii) natural and engineered systems and components that are designed or assumed to operate as barriers after permanent closure are functioning as intended and expected. The Performance Confirmation Plan provides technical information and plans for *in-situ* monitoring, laboratory and field testing, and *in-situ* experiments to carry out the objectives.

- The program started during site characterization and will continue until permanent closure.

- DOE implementation of the Performance Confirmation Program includes performance confirmation test plans that provide detailed information on the final parameters used to determine whether repository barriers are functioning as intended and expected.

- The Performance Confirmation Plan includes or cites applicable procedures for records and reports.

2.4.3.2   Confirmation of Geotechnical and Design Parameters

DOE provided information addressing confirmation of geotechnical and design parameters in SAR Sections 4.1.1, 4.1.3, 4.2.1, and 4.2.2; SAR Tables 4-1 and 4-2; and DOE (2009gm). In SAR Section 4.2.2, DOE stated that the Performance Confirmation Program includes a continuing program of surveillance, geotechnical testing, and geologic mapping to confirm geotechnical and design parameters, as well as evaluation of thermal effects on geotechnical parameters. In SAR Table 4-2, DOE identified eight performance confirmation activities and briefly described the purpose, level of current understanding, and methodology to be used for each activity in SAR Section 4.2. These activities are discussed in depth in the Performance Confirmation Plan. As described in SAR Section 4.1.1, additional details will be developed and provided in detailed performance confirmation test plans. For two of the performance confirmation activities, DOE has completed a performance confirmation test plan (BSC, 2006al; SNL, 2007bo).

The NRC staff's evaluation is organized in four sections: (i) program for measuring, testing, and geologic mapping (TER Section 2.4.3.2.1); (ii) thermomechanical response monitoring

(TER Section 2.4.3.2.2); (iii) surveillance program to evaluate subsurface conditions against design assumptions (TER Section 2.4.3.2.3); and (iv) confirmation of geotechnical and design parameters (TER Section 2.4.3.1.4).

2.4.3.2.1   Program for Measuring, Testing, and Geologic Mapping

In SAR Section 4.2.2, DOE described the geotechnical and design monitoring and testing that will occur during construction and operations to confirm geotechnical and design parameters. DOE stated that the program involves

- Identification of the geotechnical and design parameters to be monitored or measured and that specific geotechnical and design parameters planned to be measured or observed will be identified in performance confirmation test plans

- Identification of interactions between natural and engineered systems to be monitored or measured and that specific interactions between natural and engineered systems planned to be measured or observed will be identified in performance confirmation test plans

- Comparison of measurements and observations with the original design bases and assumptions

- Comparison of performance confirmation monitoring and measurement results with the original design bases and assumptions to determine the need for design modifications and construction method changes, if necessary

- Evaluation of the significance of performance confirmation monitoring and measurement results

- Reporting performance confirmation results and the evaluation of impacts on repository performance

DOE identified performance confirmation activities in SAR Table 4-2. These activities and the related SAR section in which they are described are (i) seepage monitoring (SAR Section 4.2.1.2), (ii) drift inspection (SAR Section 4.2.1.8), (iii) thermally accelerated drift near-field monitoring (SAR Section 4.2.1.9), (iv) thermally accelerated drift in-drift environment (SAR Section 4.2.1.11), (v) subsurface mapping (SAR Section 4.2.2.1), (vi) seismicity monitoring (SAR Section 4.2.2.2), (vii) construction effects monitoring (SAR Section 4.2.2.3), and (viii) thermally accelerated drift thermal-mechanical monitoring (SAR Section 4.2.2.4). For each activity, DOE described the purpose; provided a description of the current understanding, including information on baseline data; and described the methodology.

The NRC staff's evaluation begins in TER Section 2.4.3.2.1.1 with an assessment of whether, in the program for measuring, testing, and geologic mapping, (i) the geotechnical and design parameters were selected using a performance-based method and (ii) the list of selected parameters is reasonable and complete. In TER Section 2.4.3.2.1.1, the NRC staff reviews whether DOE accounted for the effects of construction, waste emplacement operations, and interactions between natural and engineered systems in its consideration of geotechnical and design parameters. The NRC staff then reviews seven of these activities in TER Sections 2.4.3.2.1.2–2.4.3.2.1.8 and reviews thermally accelerated drift thermal-mechanical monitoring in TER Section 2.4.3.2.2. In TER Section 2.4.3.2.1.9, the NRC staff summarizes its evaluation of the program for measuring, testing, and geologic mapping.

2.4.3.2.1.1   Parameter Selection

In SAR Section 4.1.1, DOE described the method of selection of the geotechnical and design parameters it will monitor and analyze and identified that it considered the need to preserve the retrieval option as an integral aspect of its Performance Confirmation Program. The NRC staff's evaluation in TER Section 2.4.3.1.1 notes that the parameters were selected using a performance-based method that focused on those parameters that could affect health and safety, because DOE used sensitivity of barrier capability and system performance to the parameter as one of the criteria in parameter selection. The NRC staff noted that the list of selected geotechnical and design parameters is reasonable and complete by comparing the results of the performance assessment to the list of selected parameters. The list of selected parameters is reasonable and complete because it is consistent with the performance assessment (e.g., seepage rates and locations are important parameters to evaluate results from the seepage model) and no geotechnical and design parameter that is important to waste isolation has been excluded. As TER Section 2.4.3.1.1 states, DOE's selection process for Performance Confirmation Program activities uses an approach consistent with identification of monitoring and testing activities for systems, components, parameters, and effects that account for the effects of construction, waste emplacement operations, and interactions between natural and engineered systems included in the design bases.

The NRC staff's review of the seepage monitoring, drift inspection, thermally accelerated drift near-field monitoring, thermally accelerated drift in-drift environment, subsurface mapping, seismicity monitoring, and construction effects monitoring performance confirmation activities focused on the selected parameters and is presented in TER Sections 2.4.3.2.1.1–2.4.3.2.1.7, respectively. The NRC staff evaluates (i) the parameters; (ii) the method to establish a baseline for the parameters; (iii) the baseline for the parameters, if the baseline has been established; and (iv) the monitoring, testing, or experimental methods for the parameters. As TER Section 2.4.3.1.1 describes, the NRC staff's evaluation focuses on the candidate parameters, except for the parameters associated with the seismicity monitoring and construction effects monitoring activities where confirmation test plans have been completed.

2.4.3.2.1.2   Seepage Monitoring

DOE described seepage monitoring in SAR Section 4.2.1.2 and Performance Confirmation Plan Section 3.3.1.2. DOE intends to use the results from the seepage monitoring, in conjunction with general monitoring for seepage in nonemplacement drifts under drift inspection, to evaluate the spatial and temporal distribution of seepage into drifts, and (if possible) to sample the chemistry of seeping water, under ambient and thermally loaded conditions. DOE stated that it plans to conduct specific tests in unventilated alcoves (using remote video systems to identify seepage) or boreholes, as early as practicable. Seepage monitoring in sealed ambient condition alcoves and in the Enhanced Characterization of the Repository Block Cross–Drift began during site characterization and will continue to be expanded to new areas through closure. DOE also stated that it plans to conduct specific tests through closure in a thermally accelerated test drift, using remote video systems to identify seepage, and to conduct humidity and temperature monitoring of the exit air at a location with suitable access and utilities that could be used to detect marked humidity or temperature changes.

In SAR Table 4-1 and Performance Confirmation Plan Table 3-2, DOE identified seepage rate, locations, and quantity and chemical composition as candidate monitoring parameters. In addition, DOE stated it would monitor the barometric pressure, temperature, and humidity of ventilated air as candidate parameters. The NRC staff notes that these candidate parameters are reasonable because they include direct measures of the quantity and quality of waters that may seep into mined openings at Yucca Mountain under ambient (unventilated) conditions, as well as indirect measures (barometric pressure, temperature, and humidity) that may be used to perform a liquid vapor water balance and thereby estimate alterations in seepage as induced by evaporation under thermally loaded conditions. The method to establish a baseline for the parameters is reasonable because baseline information will be synthesized from performance assessment results, as well as from information in analysis and model reports, and the baseline will be presented in the detailed performance confirmation test plan. DOE's proposed methodology for ambient conditions is suitable for the parameters because the methodology is consistent with previous DOE use at Yucca Mountain since site characterization. The NRC staff notes that the proposed methodology for thermally perturbed conditions is reasonable for the parameters because some of the methods (temperature and humidity monitoring) are consistent with previous DOE use at Yucca Mountain since site characterization.

2.4.3.2.1.3 Drift Inspection

DOE described the drift inspection activity in SAR Section 4.2.1.8 and Performance Confirmation Plan Section 3.3.1.8. DOE stated that the purpose of the activity is to evaluate drift stability assumptions and rockfall size, to confirm that Engineered Barrier System (EBS) components will endure, and to confirm by direct observation that the design preserves the option to retrieve waste. DOE stated that it plans to perform regular nonemplacement drift inspections, as well as periodic inspections of selected emplacement drifts and a thermally accelerated drift. DOE intends to perform reactive observations of the condition of underground openings subsequent to any significant seismic events. DOE stated that the drift inspection activity will be initiated during operations and continue until permanent closure. DOE also stated that at present certain plans are conceptual in nature because the monitoring devices are not presently available and further development of specific monitoring devices will be needed. For example, DOE explained that a remote monitoring device capable of obtaining data in the high-temperature and high-radiation environment of emplacement and thermally accelerated drifts is presently not available.

In SAR Table 4-1 and Performance Confirmation Plan Table 3-2, DOE identified temperature (as a surrogate indicator of evaporating seepage), humidity, seepage, rockfall size and frequency, ground support conditions, engineered barrier component positions (such as waste package positions and rail alignment), and drift continuity as candidate monitoring parameters. The NRC staff notes that these candidate parameters are reasonable because they include (i) direct measures of drift mechanical conditions for emplacement, thermally accelerated, and nonemplacement drifts; (ii) parameters used in drift stability calculations (e.g., temperature); and (iii) measures that can be used to evaluate the consequence of drift degradation (e.g., temperature, humidity, and seepage). The method to establish a baseline for the parameters is reasonable because DOE stated baseline information will be developed from analysis and model reports and the baseline will be presented in the detailed performance confirmation test plan. The NRC staff notes that the proposed methodology is suitable for the parameters because parts of the methodology (i.e., observation of nonemplacement drifts) have been previously used by DOE at Yucca Mountain and remote measurement in emplacement and thermally accelerated drifts is reasonable and conceptually feasible (i.e., similar inspection techniques were used in the Exploratory Studies Facility Drift Scale Test).

2.4.3.2.1.4   Thermally Accelerated Drift Near-Field Monitoring

DOE described thermally accelerated drift near-field monitoring in SAR Section 4.2.1.9 and Performance Confirmation Plan Section 3.3.1.9. DOE stated that the purpose of the near-field monitoring for the thermally accelerated drift is to provide information to evaluate results from the thermal-hydrologic-mechanical-chemical model for the near-field environment (coupled processes) to assess the modeled repository performance bases pertaining to drift seepage. DOE stated that it plans to perform monitoring using boreholes drilled into the near-field rock from an observation drift adjacent to the thermally accelerated drift, to perform monitoring using core collected from the boreholes to confirm *in-situ* rock moisture content and chemistry, and to monitor changes resulting from heating due to emplaced waste. DOE also stated that arrays of boreholes will be designed for specific tests to be performed. This requires construction of the drift to be used for the thermally accelerated tests and emplacement of actual waste packages. DOE stated that it plans to begin during operations and continue until permanent closure.

In SAR Table 4-1 and Performance Confirmation Plan Table 3-2, DOE identified rock-mass moisture content, temperature and thermal gradients, fracture air permeability, mechanical deformation, mechanical properties, and water chemistry as candidate monitoring parameters. The NRC staff notes that these candidate parameters are reasonable because thermal characteristics (i.e., temperature and thermal gradients), hydrologic characteristics (i.e., rock-mass moisture content, air permeability), mechanical characteristics (i.e., mechanical deformation and mechanical properties), and chemical characteristics (i.e., water chemistry) are the parameters important to evaluating coupled process models. DOE's method to establish a baseline for the parameters is reasonable because DOE stated that baseline information, although not yet available because it relates to conditions created by actual waste packages or to conditions in rock units different from those encountered in previous tests done during site characterization, will be developed from numerical models and acquired from thermal results reports. The NRC staff notes that the proposed methodology is reasonable for the parameters because the methodology is consistent with the techniques used since site characterization and will collect applicable information to confirm DOE assumptions and parameters used in the TSPA.

2.4.3.2.1.5   Thermally Accelerated Drift In-Drift Environment Monitoring

DOE described thermally accelerated drift in-drift environment monitoring in SAR Section 4.2.1.11, DOE (2009gm), and Performance Confirmation Plan Section 3.3.1.11. The purpose of thermally accelerated drift in-drift monitoring is to provide information to evaluate the in-drift physical and chemical environment to support evaluating performance lifetimes of the waste package container and drip shield supports. DOE stated that information it obtains will be used to evaluate assumptions used in in-drift physical and chemical environment models. DOE also stated that monitoring will begin after constructing the thermally accelerated drift and emplacing waste packages and continue until permanent closure. DOE stated that it plans to use a remote monitoring device to obtain data within the thermally accelerated drift. As demonstrated in the Drift Scale Test during site characterization, the technology to provide a remote means to make measurements in bulkheaded alcoves is available. The high-temperature and high-radiation environments representative of post emplacement conditions in a thermally accelerated drift require integration of specific technology to accomplish measurements and inspections. DOE stated that details of the test methodology will be developed and documented in the detailed performance confirmation test plan.

In SAR Table 4-1 and Performance Confirmation Plan Table 3-2, DOE identified the following candidate monitoring parameters: temperature; relative humidity; gas composition; radionuclides; pressure; radiolysis; thin films evaluation; condensation water quantities; and composition or ionic characteristics, including microbial effects. The NRC staff notes that these candidate parameters are reasonable because they directly assess the in-drift environmental conditions that affect the performance of EBS components. DOE's method to establish a baseline for the parameters is reasonable because DOE stated the expected conditions (i.e., the baseline) would be synthesized from information in analysis and model reports (note: the actual conditions created by emplaced waste packages do not exist at this time). This information will be provided in the detailed performance confirmation test plan. The NRC staff notes that the proposed methodology is suitable for the parameters because remote measurement in an emplacement drift is applicable and conceptually feasible; similar inspection techniques have been previously used by DOE at Yucca Mountain in the Exploratory Studies Facility Drift Scale Test.

2.4.3.2.1.6   Subsurface Mapping

DOE described the subsurface mapping activity in SAR Section 4.2.2.1 and in the Performance Confirmation Plan Section 3.3.2.1. The purpose of the subsurface mapping is to confirm the actual subsurface conditions encountered during construction. Underground geologic mapping ensures that observed variations from the expected geologic conditions described in the SAR are documented and provides the basis to evaluate the information on the geologic framework that was used to model and evaluate the performance of the natural systems of the repository. DOE stated that mapping will begin soon after underground construction begins; will be conducted more or less continuously as new drifts, mains, and shafts are mined; and will end soon after the last opening is mined.

In SAR Table 4-1 and Performance Confirmation Plan Table 3-2, DOE identified fracture characteristics (e.g., orientation, length, infilling, aperture), fault zone characteristics (offset, location, age), stratigraphic contacts, and lithophysal characteristics as the candidate parameters. The NRC staff notes that these candidate parameters are reasonable because they can be used directly to confirm anticipated subsurface conditions that have been used to construct models pertinent to performance assessment and repository design. The NRC staff also notes that these parameters are reasonable because they can be used to (i) confirm reasonableness of designed ground support components, (ii) evaluate short- and long-term stability of emplacement and non-emplacement openings, (iii) evaluate predictions of thermal loading on the walls of emplacement drifts, (iv) assess near-field hydrologic characteristics of the emplacement drifts, and (v) detect the presence of anomalous infillings that might have deleterious effects on waste isolation characteristics of the repository. DOE's method to establish a baseline for the parameters is reasonable because baseline data will be (i) derived from the integrated site model, with details of mapping not captured at the scale of the drifts and mains supplemented from results in the previous geologic mapping in the Exploratory Studies Facility and Enhanced Characterization of the Repository Block Cross-Drift, and (ii) presented in the detailed performance confirmation test plan for subsurface mapping. The NRC staff notes that the proposed methodology for subsurface mapping is reasonable because these techniques have been previously used by DOE at Yucca Mountain.

2.4.3.2.1.7   Seismicity Monitoring

DOE described seismicity monitoring in SAR Section 4.2.2.2, DOE (2009gm), Performance Confirmation Plan Section 3.3.2.2, and Performance Confirmation Test Plan for Seismicity

Monitoring (SNL, 2007bo). The purposes of seismicity monitoring are to assess the regional seismic activity that is used in simulations of the seismic disruption scenario that are relevant to evaluation of the Engineered Barrier System (EBS) and to collect field observations of any large magnitude fault displacement after significant local or regional seismic events. DOE stated that the methodology would include use of the existing area/regional seismic monitoring network and onsite monitoring and inspections identified in SNL Tables 2-3 and 2-4 (2007bo) following a significant seismic event. Seismicity monitoring began during site characterization, and DOE stated that it anticipates that the monitoring system will be maintained through repository closure.

DOE identified candidate parameters in SAR Table 4-1 and Performance Confirmation Plan Table 3-2, and the same candidate parameters were identified as parameters in the Performance Confirmation Test Plan for Seismicity Monitoring. In response to an NRC staff request for additional information (RAI), DOE (2009gm) clarified that test plans specify parameters that will meet the objectives of the activity and that the seismicity monitoring test plan (SNL, 2007bo) is complete; thus, the monitoring parameters are event detection, event magnitude, event location, strong-motion data collection and analysis, and seismic attenuation investigations {within 50 km [31.1 mi]}. The NRC staff notes that these parameters are reasonable because they can be used to (i) evaluate the historic earthquake information distribution and spectra, (ii) evaluate inputs for the preclosure seismic design of the repository, and (iii) assess the impact of future seismic activity on the repository. DOE's methodology to establish a baseline for the parameters is reasonable because a catalog of historical and instrumentally recorded earthquakes that was compiled for the region within 300 km [186.4 mi] of the repository site is the basis for developing the baseline for the monitored parameters. The baseline for parameters in SNL Table 2-3 (2007bo) is reasonable because it is consistent with information used to develop the seismic analysis in the SAR (BSC, 2004aj,al) and it is consistent with the Technical Work Plan for: Construction Effects Monitoring (BSC, 2006al). The NRC staff notes that the methodology is reasonable because it is consistent with the techniques used since site characterization and will collect applicable information to confirm DOE assumptions and parameters used in the TSPA.

2.4.3.2.1.8   Construction Effects Monitoring

DOE described construction effects monitoring in SAR Section 4.2.2.3, DOE (2009gm), Performance Confirmation Plan Section 3.3.2.3, and BSC (2006al). The purpose of the construction effects monitoring activity is to monitor the response of emplacement and main drift excavations to confirm drift degradation analysis predictions and assumptions, underground opening stability, and the ability to retrieve waste. Construction effects monitoring began during site characterization, and DOE stated that it expects to continue this monitoring until emplacement of waste and closure of mains and shafts. DOE's methodology includes convergence pins, multipoint extensometers, and single-point extensometers. DOE described the methods, locations, and timing of its measurements and the types of information to be collected (BSC, 2006al).

Although DOE identified them as candidate parameters in SAR Table 4-1 and Performance Confirmation Plan Table 3-2, DOE (2009gm) clarified, in response to an NRC staff RAI, that test plans specify parameters that will meet the objectives of the activity and that the construction effects monitoring test plan (BSC, 2006al) is complete; thus, the monitoring parameters are drift convergence using tape and rod extensometers, tunnel stability using visual observations, engineered ground support systems using visual observation, and geotechnical parameters (i.e., *in-situ* rock stress measurements) at selected locations. The NRC staff notes that these parameters are reasonable because they directly assess the mechanical deformation and degradation of the drift. DOE's methodology to establish a baseline for the parameters is reasonable because the baseline will be synthesized from information obtained from reports, mainly BSC (2004al), that supported the SAR and from design specifications. The baseline for parameters in BSC Tables 1-1 and 1-2 (2006al) is reasonable because it is consistent with DOE's ground control analyses that support the subsurface design. The NRC staff notes that the methodology is suitable for the parameters because these techniques have been previously used by DOE at Yucca Mountain.

2.4.3.2.1.9   Summary of the NRC Staff Evaluation of Program for Measuring, Testing, and Geologic Mapping

On the basis of the review in TER Sections 2.4.3.2.1.1–2.4.3.2.1.8, the NRC staff notes that DOE's program for measuring, testing, and geologic mapping is reasonable because of the following:

- The Performance Confirmation Plan establishes a program for measuring, testing, and geologic mapping to confirm geotechnical and design parameters, including natural processes, pertaining to natural systems and components that are assumed to operate as barriers after permanent closure.

- DOE stated that it will implement the program during repository construction and operation.

- DOE stated that the geotechnical and design parameters it will monitor and analyze were selected using a performance-based method that focused on those parameters that could affect health and safety. DOE also considered (i) the need to preserve the retrieval option, (ii) the effects of construction and waste emplacement operations, and (iii) interactions between natural and engineered systems that were evaluated in the original design bases and assumptions for the geotechnical and design parameters.

- The list of selected geotechnical and design parameters is reasonable and DOE has not excluded any geotechnical and design parameter that is important to waste isolation.

- For the seepage monitoring, drift inspection, thermally accelerated drift near-field monitoring, thermally accelerated drift in-drift environment, subsurface mapping, seismicity monitoring, and construction effects monitoring performance confirmation activities, DOE has (i) reasonable parameters or candidate parameters; (ii) a reasonable method to establish a baseline for the parameters; (iii) a reasonable baseline for the parameters for which the baseline has been established; and (iv) suitable monitoring, testing, or experimental methods for the parameters.

2.4.3.2.2   Thermomechanical Response Monitoring

This section describes the NRC staff review of DOE's plans to monitor, *in-situ*, the thermomechanical response of the underground facility until permanent closure. As described in TER Section 2.4.3.2.1.1 the NRC staff notes that (i) the method to select parameters was a performance-based method that focused on those parameters that could affect health and safety and considered the need to preserve the retrieval option; (ii) the list of selected geotechnical and design parameters is reasonable and complete and no geotechnical and design parameter that is important to waste isolation has been excluded; and (iii) DOE's selection process approach for performance confirmation activities was consistent with identification of monitoring and testing activities for systems, components, and parameters that accounted for the effects of construction, waste emplacement operations, and interactions between natural and engineered systems.

DOE described thermally accelerated drift thermal-mechanical monitoring in SAR Section 4.2.2.4, DOE (2009gm), and the Performance Confirmation Plan Section 3.3.2.4. The purpose of thermally accelerated drift thermal-mechanical monitoring is to assess drift degradation assumptions under thermal conditions, thereby providing an indication of overall drift stability, in conjunction with construction effects monitoring and drift inspection. DOE stated that it plans to remotely monitor deformation of the drift periphery and invert, anticipating that existing technology must be adapted to the high-temperature and high-radiation environment within the thermally accelerated drift. This activity will begin after construction of the thermally accelerated drift and emplacement of waste packages. DOE stated that details of test methodology will be developed and documented in the detailed performance confirmation test plan prior to waste operations.

In SAR Table 4-1 and Performance Confirmation Plan Table 3-2, DOE identified drift convergence, drift shape, drift degradation, ground support visual condition, rail alignment, invert visual condition, pallet visual condition, waste package alignment, and spacing as candidate monitoring parameters. The NRC staff notes that these candidate parameters are reasonable because they directly assess the mechanical deformation of the drift and emplaced Engineered Barrier System (EBS) components. DOE's method to establish a baseline for the parameters is reasonable because DOE stated baseline information, although not yet available because it relates to conditions created by actual waste packages, would be synthesized from ranges and distributions for geotechnical parameters used in preclosure design and postclosure analyses.

2.4.3.2.3   Surveillance Program To Evaluate Subsurface Conditions Against Design Assumptions

In SAR Section 4.2.2, DOE described that the Performance Confirmation Program includes a continuing program of surveillance, geotechnical testing, and geologic mapping to confirm geotechnical and design parameters. DOE stated that geotechnical and design monitoring and testing will occur during construction and operations. DOE also stated that the monitoring and testing program will include (i) monitoring of subsurface conditions; (ii) comparison of measurements and observations with the original design bases and assumptions; (iii) comparison of performance confirmation monitoring and measurement results with the original design bases and assumptions to determine the need for design modifications and construction method changes, if necessary, as described in SAR Section 4.1.3; (iv) evaluation of the significance of performance confirmation monitoring and measurement results, as described in SAR Section 4.1.3; and (v) reporting performance confirmation results and the evaluation of impacts on repository performance. As described in SAR Section 4.1.1, the evaluations of impacts on repository performance would provide information that could lead DOE to report recommended design or construction method changes to NRC.

The NRC staff evaluated the information in SAR Sections 4.1.1, 4.1.3, 4.2.1.8, and 4.2.2–4.2.2.4 and the Performance Confirmation Plan on DOE's surveillance program to evaluate subsurface conditions against design assumptions. The drift inspection, subsurface mapping, and construction effects monitoring performance confirmation activities constitute a surveillance program to evaluate subsurface conditions against design assumptions during construction. The NRC staff notes that the construction effects monitoring and the thermally accelerated drift thermal-mechanical monitoring performance confirmation activities constitute a surveillance program to evaluate subsurface conditions against design assumptions during operations because the information on the activities in the Performance Confirmation Plan is consistent with the geotechnical and design parameters in the SAR.

The NRC staff notes that the surveillance program is reasonable for confirmation of geotechnical and design parameters because DOE stated that the performance confirmation activities compare measurements and observations with original design bases and assumptions. DOE stated that the subsurface geologic mapping activity begins with construction, continues as new underground opening are exposed, ensures that observed variations from the expected geologic conditions described in the SAR are documented, and provides the basis to evaluate the information of the geologic framework model (e.g., thicknesses of geologic units based on stratigraphic contacts) that was used by DOE to develop the unsaturated zone flow models.

In SAR Section 4.1.3, DOE described plans to follow a procedure that compares performance confirmation results with condition limits established in the performance confirmation test plans. DOE stated it plans to (i) conduct regular integration reviews to evaluate overall performance confirmation results, with the integration review reported to NRC in a manner consistent with overall performance confirmation reporting schedule; (ii) notify NRC initially when performance confirmation results exceed condition limits established in the performance confirmation test plans; (iii) submit a subsequent evaluation report providing detailed information on the event, including recommended changes, after initial NRC notification; and (iv) provide information with respect to changes, tests, experiments and deficiencies. DOE's surveillance program is reasonable for confirming geotechnical and design parameters because it includes provisions for (i) determining the need for modifications to the design or construction methods if significant differences exist between measurements and observations and original design bases and assumptions and (ii) reporting significant differences between measurements and observations and the original design bases and assumptions, their significance to health and safety, and recommended changes to NRC.

2.4.3.2.4   Summary of the NRC Staff Evaluation on Confirmation of Geotechnical and Design Parameters

On the basis of the NRC staff's evaluation in TER Sections 2.4.3.2.1–2.4.3.2.3, the NRC staff notes that DOE's plans for confirmation of geotechnical and design parameters are reasonable because of the following:

- The Performance Confirmation Plan establishes a program for measuring, testing, and geologic mapping to confirm geotechnical and design parameters, including natural processes, pertaining to natural systems and components that are assumed to operate as barriers after permanent closure. DOE can implement the program during repository construction and operation.

- The Performance Confirmation Program to confirm geotechnical and design parameters includes reasonable plans to monitor, *in-situ*, the thermomechanical response of the underground facility until permanent closure.

- The Performance Confirmation Plan sets up a surveillance program to evaluate subsurface conditions against design assumptions.

2.4.3.3   Design Testing

DOE provided information addressing a program for testing of engineered systems and components used in the design in SAR Sections 4.1.1, 4.1.3, and 4.2.3; SAR Tables 4-1 and 4-2; and DOE (2010ap). In SAR Section 4.2.3, DOE stated that the Performance Confirmation Program for testing engineered systems and components used in the design will be developed and initiated as early as practicable during construction and will continue into the operational period. In SAR Table 4-2, DOE identified five performance confirmation activities (seepage monitoring, thermally accelerated drift near-field monitoring, construction effects monitoring, thermally accelerated drift thermal-mechanical monitoring, and seal and backfill testing) for the testing of engineered systems and components used in the design. The NRC staff evaluated the seepage monitoring, thermally accelerated drift near-field monitoring, construction effects monitoring, thermally accelerated drift thermal-mechanical monitoring, and thermally accelerated drift in-drift environment monitoring activities in TER Section 2.4.3.3.1. This TER section evaluates the information on the seal and backfill testing activity.

DOE described the seal and backfill testing activity in SAR Section 4.2.3.1, DOE (2010ap), and Performance Confirmation Plan Section 3.3.3.1. The purpose of the seal and backfill activity is to evaluate design assumptions for implementing methods to seal (backfilling and plugging) shafts, ramps, and boreholes for permanent repository closure. DOE stated that it plans to backfill shafts, ramps, and boreholes. Boreholes would then be capped (sealed). DOE also stated that it will develop a final design for shaft and ramp seals on the basis of information gained from construction measurements and observations. In SAR Section 4.2.3.1, DOE stated that seal and backfill testing includes laboratory testing of the effectiveness of borehole seals, field testing of the effectiveness of ramp and shaft seals, and field testing of backfill placement and compaction procedures. DOE stated that laboratory tests on borehole seals will be completed early during construction and that field tests will be completed prior to backfilling shafts and ramps, or backfilling and plugging boreholes, and prior to permanent closure.

In SAR Table 4-1 and Performance Confirmation Plan Table 3-2, DOE identified the configuration and performance of shaft, ramp, and borehole seal materials, and the laboratory and field hydraulic and pneumatic seal effective permeability as the candidate test parameters. The NRC staff notes that these candidate parameters are reasonable because they include (i) evaluations of the seal materials and performance and (ii) parameters that directly measure seal performance (e.g., hydraulic and pneumatic effective permeability). DOE's method to establish a baseline for the parameters is reasonable because (i) DOE plans to develop baseline information for the performance, testing, analysis, placement, and compaction of backfill materials, using information in engineering literature, and (ii) DOE stated that baseline data will be presented in the detailed performance confirmation test plan for seal and backfill testing. The NRC staff notes that the proposed methodology is suitable for the parameters because information for the performance, testing, analysis, placement, and compaction of backfill materials used to seal the shafts, ramps, and boreholes is widely available in the engineering literature.

DOE's program of tests to evaluate the effectiveness of borehole, shaft, and ramp seals before full-scale sealing is reasonable because (i) the program of tests is consistent with techniques in the engineering literature; (ii) these components (borehole, shafts, and ramps) are not important to waste isolation; and (iii) the program of tests addresses design testing. The NRC staff notes that DOE determined the boreholes, shafts, and ramps are not important to waste isolation. The NRC staff notes that the program of tests is reasonable, because the proposed methodologies are consistent with engineering practice for sealing of boreholes, shafts, and ramps. DOE considered the effects of waste emplacement operations, and the program of tests to evaluate the effectiveness of borehole, shaft, and ramp seals, before full-scale sealing, on interactions between the natural and engineered systems in estimates of the intended and anticipated design bases.

2.4.3.4 Monitoring and Testing Waste Packages

DOE provided information addressing monitoring and testing of the waste package in SAR Sections 4.1, 4.1.1, 4.1.3, 4.2, and 4.2.4; SAR Tables 4-1 and 4-2; and DOE (2010ap, 2009gm). In SAR Section 4.2.4, DOE stated that performance confirmation activities include remote monitoring of a representative set of waste packages, as well as provisions for laboratory testing of waste package and drip shield materials representative of those emplaced in the repository, and that these performance confirmation activities focus on corrosion testing of waste package and drip shield materials, as well as internal waste package conditions. In SAR Table 4-2, DOE identified six performance confirmation activities (dust buildup monitoring, thermally accelerated drift in-drift environment monitoring, waste package monitoring, corrosion testing, corrosion testing of thermally accelerated drift samples, and waste form testing). DOE briefly described the purpose, level of current understanding, and methodology to be used for each activity in SAR Section 4.2. These activities are discussed in depth in the Performance Confirmation Plan. As described in SAR Section 4.1.1, additional details will be developed and provided in detailed performance confirmation test plans.

The NRC staff's evaluation is in the following four sections: (i) program for monitoring and testing the condition of waste packages (TER Section 2.4.3.4.1), (ii) program of laboratory experiments that focuses on the internal condition of the waste packages (TER Section 2.4.3.4.2), (iii) schedule for the waste package program (TER Section 2.4.3.4.3), and (iv) NRC staff evaluation on monitoring and testing waste packages (TER Section 2.4.3.4.4). In TER Section 2.4.3.4.1 the NRC staff evaluates the dust buildup monitoring, thermally accelerated drift in-drift environment monitoring, waste package monitoring, corrosion testing, and corrosion testing of thermally accelerated drift samples performance confirmation activities. In TER Section 2.4.3.4.2 the NRC staff evaluates the information on the waste form testing activity.

2.4.3.4.1   Program for Monitoring and Testing the Condition of Waste Packages

In SAR Section 4.2.4 and Performance Confirmation Plan Section 3.3.4, DOE described its program for monitoring and testing the condition of the waste packages. DOE identified that dust buildup monitoring and thermally accelerated drift in-drift environment monitoring activities are designed, in part, to provide for the monitoring of emplaced waste packages. These waste packages are representative in terms of materials, design, structure, fabrication, inspection methods, and environment, and the environmental conditions monitored in these activities address coupled thermal-hydrologic-chemical processes that affect the amount and chemistry of the water and other environmental variables. DOE described that the performance confirmation activities in the program for monitoring and testing the condition of the waste package focus on corrosion testing of waste package and drip shield materials, as well as internal waste package conditions. In SAR Section 4.2.4, DOE stated that, to the extent practicable, the laboratory experiments will be designed to include representative repository emplacement environments. DOE expects that these laboratory tests will confirm the design basis for the waste package and confirm information that supports modeling of waste package and drip shield performance. Waste package monitoring includes corrosion monitoring, but is not limited to the use of corrosion coupons.

The NRC staff's review of dust buildup monitoring, thermally accelerated drift in-drift environment monitoring, waste package monitoring, corrosion testing, and corrosion testing of thermally accelerated drift samples confirmation activities relative to the selected parameters is presented in TER Sections 2.4.3.4.1.1–2.4.3.4.1.5, respectively. For each activity, the NRC staff evaluates (i) the candidate parameters; (ii) the method to establish a baseline for the candidate parameters; and (iii) the monitoring, testing, or experimental methods for the candidate parameters. The NRC staff summarizes its evaluation of these activities in TER Section 2.4.3.4.1.6.

2.4.3.4.1.1   Dust Buildup Monitoring

DOE described the dust buildup monitoring activity in SAR Section 4.2.1.10 and Performance Confirmation Plan Section 3.3.1.10. The purpose of the activity is to evaluate assumptions of dust buildup and potential chemical effects on EBS components (waste package and drip shield). DOE stated that dust buildup contributes to corrosion of the waste package and the drip shield because of possible impacts on water chemistry and deliquescence, and that this activity is important to evaluating the bases for the expected conditions and assessing whether the environments being used in the waste package and waste form testing are representative. DOE stated that it plans to collect and analyze waste package and drip shield material specimens exposed in emplacement drifts to measure salts present in the dust and plans to collect dust from the thermally accelerated drift and other selected locations. DOE described a conceptual plan to use a remotely operated vehicle to collect dust samples, with periodic dust collection occurring during operations until closure. The details for this performance confirmation activity, as well as the details on remote methods for sample collection, will be developed, finalized, and documented in the performance confirmation test plan.

In SAR Table 4-1 and Performance Confirmation Plan Table 3-2, DOE identified quantity, physical properties, and chemical composition of dust deposited on waste package, drip shield, rail, and ground support surfaces as candidate monitoring parameters. The NRC staff notes that these candidate parameters are reasonable because they include direct measures of the physical properties and chemical composition of dusts that may accumulate on the surfaces of the waste package and drip shield and because that information is used in the evaluation of the corrosion of Engineered Barrier System (EBS) components in the performance assessment. DOE's method to establish a baseline for the candidate parameters is reasonable because (i) baseline information and expected variability will be developed from the analysis and model reports on in-drift physical and chemical environments and (ii) the baseline information will be presented in the detailed performance confirmation test plan. The NRC staff notes that the proposed methodology is reasonable because waste package and drip shield material specimens exposed in emplacement drifts will be collected and analyzed to measure the actual salts that are present in the dust.

Because DOE will be collecting samples from emplacement drifts, the NRC staff notes that the environment of the waste packages DOE will monitor and test is representative of the emplacement environment.

2.4.3.4.1.2 Thermally Accelerated Drift In-Drift Environment Monitoring

In TER Section 2.4.3.2.1.5, the NRC staff notes the candidate parameters for the thermally accelerated drift in-drift environment monitoring activity are applicable; the methodology for establishing a baseline is reasonable; and DOE has identified reasonable monitoring, testing, or experimental methods for the parameters. The NRC staff evaluation in this section focuses on the thermally accelerated drift in-drift environment relative to monitoring and testing the condition of waste packages.

The NRC staff notes that the environmental conditions DOE will monitor and evaluate are reasonable because (i) DOE identified in SAR Section 4.2.1.11 that the activity includes monitoring and laboratory evaluation of gas composition; water quantities, composition, and ionic characteristics (including thin films); microbial types and amounts; and radiation and radiolysis effects within a thermally accelerated drift and (ii) the candidate parameters include temperature and relative humidity. Because DOE described in SAR Section 4.2.1.11 that confirming the environment that surrounds the waste package container and drip shield supports evaluating the performance lifetimes of Engineered Barrier System (EBS) components, the thermally accelerated drift in-drift environment monitoring activity is reasonable for confirming performance assessment models and assumptions.

2.4.3.4.1.3 Waste Package Monitoring

DOE described the waste package monitoring activity in SAR Section 4.2.4.1 and Performance Confirmation Plan Section 3.3.4.1. The purpose of waste package monitoring is to (i) confirm the condition of selected, representative waste packages in the repository emplacement drifts; (ii) evaluate waste package integrity; and (iii) confirm the absence of leakage and leak paths. DOE stated that field waste package monitoring, including the number of packages monitored; locations, durations, and design of the testing; and waste packages selected for underground monitoring, will represent those to be emplaced in terms of materials, design, structure, fabrication, and inspection methods. In addition, waste packages selected for monitoring, on a yearly basis or possibly less frequently, will be representative of different configurations, as well as different environmental conditions that may occur in the underground facility. DOE described a conceptual plan to use remote visual observation to monitor the selected set of waste packages for external corrosion evidence and indicated the technology for a remote means to monitor waste packages is available. DOE stated that performance confirmation monitoring will be integrated with underground operations to develop a remotely operated vehicle or other monitoring technology compatible with operations. The high-temperature and high-radiation environments representative of post emplacement conditions require integration of specific technology applications to accomplish measurement and inspections. DOE stated that it is also considering the potential for using technology to sense the differential pressure between the waste package inner and outer sections. DOE indicated this activity is still preliminary in nature and other testing methods and approaches may be employed and will be documented in the detailed performance confirmation test plan. DOE stated that it plans to initiate waste package monitoring during the early stages of waste emplacement operations and will continue as long as practical until permanent closure.

In SAR Table 4-1 and Performance Confirmation Plan Table 3-2, DOE identified external visual corrosion and possibly internal pressure of the waste package as candidate monitoring parameters. These candidate parameters are reasonable because they include direct and indirect measures of waste package corrosion under relevant repository environmental conditions. The method to establish a baseline for the candidate parameters is also reasonable, because DOE stated that baseline data for this activity will be synthesized from performance assessment results, as well as from information obtained from analysis and model reports, and the baseline information will be presented in the detailed performance confirmation test plan. The NRC staff notes that the proposed methodology is reasonable because remote observation is feasible.

2.4.3.4.1.4 Corrosion Testing

DOE described corrosion testing in SAR Section 4.2.4.2 and Performance Confirmation Plan Section 3.3.4.2. The purpose of the corrosion testing activity is to confirm information used to evaluate the performance of the materials that will be used to fabricate the waste package, emplacement pallet, and drip shield components of the Engineered Barrier System (EBS). DOE stated that it will measure the general corrosion rate and evaluate the overall corrosion performance of the waste package outer barrier material and the drip shield material to assess results of corrosion models used in the performance assessment. Long-term corrosion tests, thermal aging tests, and electrochemical testing comprise the laboratory testing. DOE-planned tests for Alloy 22, Type 316 stainless steel, and titanium alloys include (i) continuation of tests, using corrosion coupons, done at the Long-Term Corrosion Test Facility to collect data on general corrosion rates, passive film properties, localized corrosion and stress corrosion cracking susceptibility, and posttest sample corrosion features; (ii) thermal aging tests to evaluate phase transformations in Alloy 22; and (iii) shorter term, electrochemically based testing for obtaining parameters in predicting localized corrosion susceptibility and measuring general corrosion rates. DOE states that the samples will be representative of the waste package, waste package pallet, and drip shield materials, including processes used to fabricate, assemble, weld, and stress relieve these materials. Corrosion testing started during site characterization, and DOE stated that it will continue until permanent closure.

In SAR Table 4-1 and Performance Confirmation Plan Table 3-2, DOE identified measurements of Alloy 22, Stainless Steel Type 316L, and Titanium Grade 7 and Grade 29 mass loss rate, passive current density, surface dissolution, open circuit potential, critical potential, stress corrosion cracking, microbial effects, surficial passive film stability, and mechanical properties as the candidate parameters. The NRC staff notes that these candidate parameters are reasonable because they directly assess the corrosion properties of the EBS materials that may be used to confirm the results of corrosion models.

DOE's method to establish a baseline for the candidate parameters is reasonable, because baseline data for this activity will be synthesized by DOE from performance assessment results, as well as from information obtained from analysis and model reports, and the baseline information will be presented in the detailed performance confirmation test plan. The NRC staff notes that the proposed methodology is reasonable because these techniques have been previously used by DOE at Yucca Mountain and most are based on American Society for Testing and Materials methods, with modification for repository-specific circumstances.

2.4.3.4.1.5 Corrosion Testing of Thermally Accelerated Drift Samples

DOE described corrosion testing of thermally accelerated drift samples activity in SAR Section 4.2.4.3 and Performance Confirmation Plan Section 3.3.4.3. The purpose of this activity is to confirm information used to evaluate corrosion models in the performance assessment. DOE plans to (i) expose test samples (often referred to as coupons), which will be representative of the waste package and drip shield materials, to the environment in the thermally accelerated drift and (ii) subsequently characterize the samples in the laboratory to measure the general corrosion rate and evaluate the corrosion performance of the waste package outer barrier material and drip shield material. For the test samples, DOE stated that specimen characteristics, postexposure characterization, and analyses will be consistent with those proposed for the laboratory corrosion testing. DOE stated that it plans to initiate this activity during operations and stated that after exposure in the thermally accelerated drifts for periods of several years, and up to the length of time before the repository is closed, samples will be periodically withdrawn for subsequent analysis and characterization in a laboratory.

In SAR Table 4-1 and Performance Confirmation Plan Table 3.2, DOE identified measurements of Alloy 22, Stainless Steel Type 316L, and Titanium Grade 7 and Grade 29 mass loss rate, passive current density, surface dissolution, open circuit potential, critical potential, stress crack corrosion, microbial effects, surficial passive film stability, and mechanical properties as the candidate parameters. The NRC staff notes these candidate parameters are reasonable because they directly assess the corrosion properties of the Engineered Barrier System (EBS) materials that may be used to confirm the results of corrosion models. DOE stated that it will synthesize baseline data for this activity from performance assessment results, as well as from information obtained from analysis and model reports, and the baseline information will be presented in the detailed performance confirmation test plan. DOE's proposed methodology is reasonable because (i) although the methods for remote emplacement and removal of test specimens are conceptual, DOE stated that further information will be provided when detailed test plans are developed and (ii) the laboratory analysis techniques have been previously used by DOE at Yucca Mountain and most are based on American Society for Testing and Materials methods.

The environment of the waste packages DOE will test is representative of the emplacement environment and consistent with safe operation because the samples, subjected to emplacement environmental conditions, will be handled remotely. The NRC staff notes that this activity's laboratory experiments are reasonable because DOE stated it will provide data needed to confirm performance assessment models and assumptions and include monitoring (via periodic withdrawal of samples and specimen characterization) and testing of corrosion coupons.

2.4.3.4.1.6   Summary of the NRC Staff Evaluation on Program for Monitoring and Testing the Condition of Waste Packages

On the basis of the NRC staff's evaluation in TER Sections 2.4.3.4.1.1–2.4.3.4.1.5, the NRC staff notes that DOE's program for monitoring and testing the condition of the waste packages is reasonable because of the following:

- The waste packages DOE will monitor and test are representative of those to be emplaced in terms of materials, design, structure, fabrication, and inspection methods.

- The environment of the waste packages DOE will monitor and test is representative of the emplacement environment and is consistent with safe operations.

- The environmental conditions DOE will monitor and evaluate include, but are not limited to, those describing water chemistry.

- Monitoring and testing include evaluation of closure welds.

- The program is technically feasible, and the sensors and devices to be used are either able to sustain the prevailing environmental conditions (e.g., temperature, humidity, radiation) during the period of repository operation or are replaceable.

- The laboratory experiments provide data needed to design the waste package and confirm performance assessment models and assumptions.

- Corrosion monitoring and testing include, but are not limited to, the use of corrosion coupons.

2.4.3.4.2　　　　　Program of Laboratory Experiments That Focus on the Internal Condition of Waste Packages

DOE described the waste form testing activity in SAR Section 4.2.4.4, Performance Confirmation Plan Section 3.3.4.4, and DOE (2009gm). The NRC staff evaluated (i) the candidate parameters; (ii) the method to establish a baseline for the parameters; and (iii) the monitoring, testing, or experimental methods for the parameters.

The purpose of the waste form testing activity is to evaluate results of waste form degradation models and evaluate in-package expected conditions. As described in SAR Section 4.2.4.4, movement of liquid or vapor-phase water through cracks in the waste package can initiate coupled processes, including degradation of fuel and steel components inside the package, and the interacting processes control the availability of water in the waste package, the pH in the package, and, by extension, the solubility and colloid mobility of dose-critical radionuclides. DOE described that the activity includes waste form testing and will include waste package coupled effects in the laboratory under simulated internal waste package conditions. DOE stated that this activity will provide information to confirm its assumptions used in the SAR for waste package source-term models used in performance assessments. DOE stated that aspects of this activity began during site characterization and will continue until at least the early stages of waste emplacement.

In SAR Table 4-1 and Performance Confirmation Plan Table 3-2, DOE identified radionuclide release rate, dissolution rate, environmental and hydrochemical indicators (Eh, pH, colloid characteristics), bare waste form dissolution, fuel rod waste form dissolution, and waste form and waste package performance under coupled chemical environment as the candidate parameters. In DOE (2009gm), DOE stated that pH-buffering capabilities of stainless steel corrosion products, aqueous chemical characteristics of the corrosion products domain, radionuclide sorption properties of stainless steel corrosion products, and colloid generation potential of corroding stainless steel under in-package conditions were captured in the environmental and hydrochemical indicators (Eh, pH, colloid characteristics) and waste form and waste package performance under coupled chemical environment general candidate parameters. DOE stated that during development of the performance confirmation test plan for waste form testing, parameters important to radionuclide releases from the waste package will be evaluated and the performance confirmation test plan for waste form testing, including the final parameter list, will be completed during construction. DOE (2009gm) stated it would provide the performance confirmation test plan to NRC prior to test implementation. The NRC staff notes the candidate parameters are reasonable because (i) they include indirect assessment of waste form dissolution under relevant repository conditions; (ii) DOE stated that the candidate parameters include pH-buffering capabilities of stainless steel corrosion products, aqueous chemical characteristics of the corrosion products domain, radionuclide sorption properties of stainless steel corrosion products, and colloid generation potential of corroding stainless steel under in-package conditions; (iii) the final parameters will be specified in the test plan for waste form testing; and (iv) DOE stated it would submit the final specific parameters to NRC prior to test implementation. The DOE baseline data for this activity will be synthesized from performance assessment results, as well as from information obtained from analysis and model reports, and the baseline information will be presented in the detailed performance confirmation test plan. The NRC staff notes the planning for this activity is conceptual at this time, and DOE stated that further details would be provided when the detailed test plan is developed.

2.4.3.4.3   Schedule for the Waste Package Program

In SAR Section 4.2.4 and DOE (2009gm), DOE described the schedule for the monitoring and testing waste package performance confirmation activities. For instance, the corrosion testing and waste form testing performance confirmation activity began during site characterization and will continue until repository closure (for the waste form activity the tests will continue until at least the early stages of waste emplacement). DOE stated that the dust buildup monitoring, thermally accelerated drift in-drift environment monitoring, waste package monitoring, and corrosion testing of thermally accelerated drift samples performance confirmation activities will be initiated during repository operations, and the different activities will terminate at different times during operations up to repository closure.

The NRC staff notes that the schedule for the waste package program is reasonable because DOE began its monitoring and testing of the waste package program as soon as practicable by beginning the corrosion testing and waste form testing activities during site characterization and other DOE activities are planned to occur once the facility for conducting the activity is available (e.g., thermally accelerated drift) and the materials (i.e., waste packages) have been emplaced. The NRC staff further notes that DOE's waste package program is reasonable because DOE stated it will continue its monitoring and testing of the waste package program as long as practicable up to permanent closure.

2.4.3.4.4   Summary of the NRC Staff Evaluation on Monitoring and Testing
                   Waste Packages

On the basis of the NRC staff's evaluation in TER Sections 2.4.3.4.1–2.4.3.4.3, the NRC staff notes that DOE's testing and monitoring of waste packages are reasonable because of the following:

•   The Performance Confirmation Plan establishes a program for monitoring and testing the condition of waste packages at the geologic repository operations area (GROA).

•   The Performance Confirmation Plan establishes a program of laboratory experiments that focuses on the internal condition of the waste packages.

•   The environment experienced by the emplaced waste packages is duplicated in the laboratory experiments to the extent practicable.

•   The schedule for the waste package program requires monitoring and testing to begin as soon as practicable, and monitoring and testing will continue as long as practical up to the time of permanent closure.

2.4.4 NRC Staff Conclusions

The NRC staff notes that DOE has provided a reasonable description of its Performance Confirmation Plan. The DOE's Performance Confirmation Plan is consistent with the guidance in the YMRP.

2.4.5 References

BSC. 2006al. "Technical Work Plan for: Construction Effects Monitoring." TWP–MGR–GE–000006. Rev. 01. ACC:DOC.20060915.0004. Las Vegas, Nevada: Bechtel SAIC Company.

BSC. 2004aj. "Development of Earthquake Ground Motion Input for Preclosure Seismic and Postclosure Performance Assessment of a Geologic Repository at YM Nevada." MDL–MGR–GS–000003. Rev. 01. ACN 01. Las Vegas, Nevada: Bechtel SAIC Company, LLC.

BSC. 2004al. "Drift Degradation Analysis." ANL–EBS–MD–000027. Rev. 03. ACN 001, ACN 002, ACN 003, ERD 01. Las Vegas, Nevada: Bechtel SAIC Company, LLC.

DOE. 2010ap. "Yucca Mountain—Response to Request for Additional Information Regarding License Application (Safety Analysis Report Section 4), Safety Evaluation Report Vol. 4, Chapter 2.4, Set 2." Letter (February 23) J.R. Williams to F. Jacobs (NRC). ML100541535. Washington, DC: DOE, Office of Technical Management.

DOE. 2009an. "Yucca Mountain—Response to Request for Additional Information Regarding License Application (Safety Analysis Report Section 2.1), Safety Evaluation Report Vol. 3, Chapter 2.2.1.1, Set 1." Letter (February 6) J.R. Williams to J.H. Sulima (NRC). ML090400455. Washington, DC: DOE, Office of Technical Management.

DOE. 2009av. DOE/RW–0573, "Yucca Mountain Repository License Application." Rev. 1. ML090700817. Las Vegas, Nevada: DOE, Office of Civilian Radioactive Waste Management.

DOE. 2009gm. "Yucca Mountain—Response to Request for Additional Information Regarding License Application (Safety Analysis Report Section 4), Safety Evaluation Report Vol. 4, Chapter 2.4, Set 1." Letter (October 28) J.R. Williams to F. Jacobs (NRC). ML093020092. Washington, DC: DOE, Office of Technical Management.

NRC. 2003aa. NUREG–1804, "Yucca Mountain Review Plan—Final Report." Rev. 2. ML032030389. Washington, DC: NRC.

NRC. 2001aa. "Disposal of High-Level Radioactive Wastes in a Proposed Geologic Repository at Yucca Mountain, Nevada: Final Rule." *Federal Register*. Vol. 66, No. 213. pp. 55732–55816. Washington, DC: NRC.

SNL. 2008ad. "Postclosure Nuclear Safety Design Bases." ANL–WIS–MD–000024. Rev. 01. ACN 01, ERD 01, ERD 02. Las Vegas, Nevada: Sandia National Laboratories.

SNL. 2008ag. "Total System Performance Assessment Model/Analysis for the License Application." MDL–WIS–PA–000005. Rev. 00. AD 01, ERD 01, ERD 02, ERD 03, ERD 04. ML090790353. Las Vegas, Nevada: Sandia National Laboratories.

SNL. 2008aq. "Performance Confirmation Plan." TDR–PCS–SE–000001. Rev. 05. ADD 01. Las Vegas, Nevada: Sandia National Laboratories.

SNL. 2007az. "Simulation of Net Infiltration for Present-Day and Potential Future Climates." MDL–NBS–HS–000023. Rev. 01. AD 01, ERD 01, ERD 02. Las Vegas, Nevada: Sandia National Laboratories.

SNL. 2007bo. "Performance Confirmation Test Plan for Seismicity Monitoring." TWP–MGR–MM–000003 Rev 00, Las Vegas, Nevada: Sandia National Laboratories. ACC: DOC.2070702.0003.

SNL. 2007bp. "Performance Confirmation Test Plan for Precipitation Monitoring." TWP–MGR–MM–000002 Rev 01, Las Vegas, Nevada: Sandia National Laboratories. ACC: DOC.20071114.0008.

CHAPTER 3

2.5.1 Quality Assurance Program

2.5.1.1　　　　Introduction

This chapter evaluates the description provided in the U.S. Department of Energy (DOE) Safety Analysis Report (SAR) Section 5.1 (DOE, 2008ab, 2009av) of the quality assurance program to be applied to the structures, systems, and components (SSCs) important to safety and to the engineered and natural barriers important to waste isolation. DOE's quality assurance program is described in the Office of Civilian Radioactive Waste Management (OCRWM) Quality Assurance Requirements and Description (QARD) (DOE, 2008af). DOE provided further information and proposed changes to the QARD and SAR in its responses (DOE, 2009gn,gr,gs, 2008aj,ak) to NRC staff requests for additional information.

2.5.1.2　　　　Evaluation Criteria

10 CFR 63.21(c)(20) requires that the SAR include a description of the quality assurance program to be applied to the SSCs important to safety and to the engineered and natural barriers important to waste isolation. 10 CFR 63.21(c)(20) also requires that the description of the quality assurance program must include a discussion of how the applicable requirements of 10 CFR 63.142 will be satisfied. 10 CFR Part 63, Subpart G (10 CFR 63.141–144) provides the requirements for quality assurance. 10 CFR 63.141 provides the requirements for the scope of quality assurance. 10 CFR 63.142 provides requirements for the description of the quality assurance program, applicability requirements for the quality assurance program, and 18 quality assurance criteria. 10 CFR 63.143 provides an implementation requirement. 10 CFR 63.144 provides the requirements for quality assurance program changes. 10 CFR 63.44 contains the requirements for changes, tests, and experiments. 10 CFR 63.73 contains the requirements for reports of deficiencies.

NRC staff reviewed DOE's description of its quality assurance program using guidance in the Yucca Mountain Review Plan (YMRP) Section 2.5.1 (NRC, 2003aa). The acceptance criteria follow:

- Organizational elements responsible for the Quality Assurance (QA) program are acceptable

- Activities related to the QA program are acceptable

- Activities related to design control are acceptable

- Activities related to procurement document control are acceptable

- Activities related to instructions, procedures, and drawings are acceptable

- Activities related to document control are acceptable

- Activities related to control of purchased material, equipment, and services are acceptable

- Activities related to identification and control of materials, parts, and components (including samples) are acceptable

- Activities related to control of special processes are acceptable

- Activities related to inspection are acceptable

- Activities related to test control are acceptable

- Activities related to control of measuring and test equipment are acceptable

- Activities related to handling, storage, and shipping are acceptable

- Activities related to inspection, test, and operating status are acceptable

- Activities related to nonconforming materials, parts, or components are acceptable

- Activities related to corrective action are acceptable

- Activities related to the QA records are acceptable

- Activities related to audits are acceptable

2.5.1.3 Technical Evaluation

DOE stated that it organized the QARD according to the 18 QA topics discussed in the YMRP. In addition, DOE's QARD has five supplements (Software, Sample Control, Scientific Investigation, Field Surveying, and Control of the Electronic Management Information) and two appendices (Waste Custodian Interface, and Storage and Transportation) addressing specific activities. DOE's QARD referenced the nuclear industry standard Quality Assurance Requirements for Nuclear Plants, NQA–1–1983 (American Society of Mechanical Engineers, 1983aa).

DOE's Quality Assurance Requirements and Descriptions (QARD) is organized consistent with the 18 QA topics discussed in the YMRP. NRC staff identified internal inconsistencies in the QARD associated with (i) use of a graded approach with respect to important to safety and important to waste isolation SSCs and related activities and (ii) whether personnel performing tests shall be trained, qualified, and certified. Also, NRC staff identified portions of the QARD that were unclear. In its response to NRC staff's requests for additional information (RAIs), DOE (2008aj,ak) stated it would update the QARD to ensure that the QARD was internally consistent. Additionally, DOE clarified in the RAI responses that (i) responsible management would document root causes of significant conditions adverse to quality and corrective actions taken to prevent recurrence, (ii) Subsection 7.2.14B.1 instead of Subsection 7.2.1B.1 would be cross referenced in QARD Section 7.2.12, (iii) characterization measurements of commercial and DOE spent nuclear fuel are not part of the Material Control and Accounting program, (iv) DOE considers deviations as conditions adverse to quality, and (v) procurement documents at all tiers identify the documentation to be submitted for information, review, or approval by the purchaser.

In the QARD, DOE identified and provided a basis for modifications taken from documents referenced in YMRP Section 2.5.1.5. Significant topics for modifications and clarifications in the QARD include:

- Use of personnel trained in quality assurance in lieu of persons from the quality assurance organization

- Organizational responsibilities assumed by the line organizations that had traditionally been assigned to the quality assurance organization, such as concurrence with procedures, dispositions, and corrective action statements

- Use of an existing data qualification method in addition to those identified in NUREG–1298 (NRC, 1988ab)

- Nondestructive examination personnel requirements that follow later versions of industry standards

- Alternative records control practices

NRC staff identified that it was unclear whether DOE retains total responsibility for assuring that entities which perform quality-affecting work [e.g., waste custodians and their contractors (QARD, Appendix A), and NRC licensees/certificate holders and their contractors (QARD, Appendix C)] comply with the applicable requirements of DOE's QARD. In response to an NRC staff RAI, DOE (2008ak) stated it would provide updates to the QARD. In the proposed change to QARD, Appendix A, DOE added, "The QARD does not apply to non-OCRWM commercial licensees/certificate holders licensed pursuant to the provisions of 10 CFR 50, 71, or 72 that manage SNF [spent nuclear fuel] and/or HLW [high-level waste] in accordance with an approved QA program." DOE explained that QA programs conducted under the provisions of 10 CFR 50, 71, or 72 provide quality assurance requirements comparable to those in the QARD.

The QARD also states, "additional quality assurance-related requirements governing the interface between commercial nuclear utilities and the OCRWM will be addressed in greater detail in future QARD revisions when OCRWM deems such an action to be necessary." Also, "if OCRWM identifies specific technical or quality assurance requirements for loading/storage of disposal canisters that need to be met by the utilities, these requirements would be addressed in the standard contract between OCRWM and commercial nuclear facilities." DOE stated that it will impose necessary quality assurance requirements on utilities through the DOE standard contract.

DOE's description of its quality assurance program is reasonable because DOE's Quality Assurance Requirements and Descriptions (QARD) describes the procedures and controls consistent with the 18 QA topics discussed in the YMRP. The QARD includes procedures and controls for the QA activities including, for example, procurement planning, document preparation, expert elicitation, peer reviews, readiness reviews, archiving samples, calibration, quality trends, qualification and certification of nondestructive examination personnel, source verification, and design verification. The NRC staff notes that DOE provided a reasonable basis for modifications taken from documents referenced in the YMRP and that these modifications are applicable controls for the activities. Additionally, DOE's description includes (i) the criteria applicable to SSCs important to safety, design and characterization of barriers important to waste isolation, and related activities; (ii) definitions and controls applicable to DOE commercial-grade item dedication and equivalent controls applicable to commercial procurement of analytical services, data, and calibration services; (iii) procedures for control of changes, tests, and experiments; and (iv) procedures for reporting deficiencies.

Finally, DOE (2008aj,ak) stated that it would update the QARD to ensure that the QARD is internally consistent and clear. DOE also stated that the QARD will be revised, as necessary, to address future activities related to facility operations, permanent closure of the repository, and decommissioning and dismantlement of the surface facilities.

2.5.1.4 NRC Staff Conclusions

The NRC staff notes that DOE's description of its quality assurance program is reasonable. DOE's description of its quality assurance program, including the QARD, is consistent with the guidance in the YMRP.

2.5.1.5 References

American Society of Mechanical Engineers. 1983aa. ANSI/ASME NQA-1-1983, "Quality Assurance Program Requirements for Nuclear Facilities." New York City, New York: American Society of Mechanical Engineers.

DOE. 2009av. DOE/RW–0573, "Yucca Mountain Repository License Application." Rev. 1. ML090700817. Las Vegas, Nevada: DOE, Office of Civilian Radioactive Waste Management.

DOE. 2009gn. "Yucca Mountain—Response to Request for Additional Information Regarding License Application (Safety Analysis Report Section 5.1), Safety Evaluation Report Vol. 4, Chapter 2.5.1, Set 1." Letter (February 6) J.R. Williams to B.J. Benney (NRC). ML090371019. Washington, DC: DOE, Office of Technical Management.

DOE. 2009gr. "Yucca Mountain—Response to Request for Additional Information Regarding License Application (Safety Analysis Report Section 5.1), Safety Evaluation Report Vol. 4, Chapter 2.5.1, Set 2." Letter (February 12) J.R. Williams to B.J. Benney (NRC). ML090430577. Washington, DC: DOE, Office of Technical Management.

DOE. 2009gs. "Yucca Mountain—Supplemental Response to Request for Additional Information Regarding License Application (Safety Analysis Report Section 5.1), Safety Evaluation Report Vol. 4, Chapter 2.5.1, Set 1." Letter (August 11) J.R. Williams to B.J. Benney (NRC). ML092360006. Washington, DC: DOE, Office of Technical Management.

DOE. 2008ab. DOE/RW–0573, "Yucca Mountain Repository License Application." Rev. 0. ML081560400. Las Vegas, Nevada: DOE, Office of Civilian Radioactive Waste Management.

DOE. 2008af. DOE/RW–0333P, "Quality Assurance Requirements and Description (QARD)." Rev. 20. ML080450334. Las Vegas, Nevada: DOE, Office of Civilian Radioactive Waste Management.

DOE. 2008aj. "Yucca Mountain—Request for Additional Information Regarding License Application (Safety Analysis Report Section 5.1), Safety Evaluation Report Vol. 4, Chapter 2.5.1, Set 1." Letter (December 2) J.R. Williams to B.J. Benney (NRC). ML083380568. Washington, DC: DOE, Office of Technical Management.

DOE. 2008ak. "Yucca Mountain—Request for Additional Information Regarding License Application (Safety Analysis Report Section 5.1), Safety Evaluation Report Vol. 4, Chapter 2.5.1, Set 1." Letter (December 10) J.R. Williams to B.J. Benney (NRC). ML090080758. Washington, DC: DOE, Office of Technical Management.

NRC. 2003aa. NUREG–1804, "Yucca Mountain Review Plan—Final Report." Rev. 2. ML032030389. Washington, DC: NRC.

NRC. 1988ab. NUREG–1298, "Qualification of Existing Data for High-Level Nuclear Waste Repositories." Washington, DC: NRC.

CHAPTER 4

2.5.2 Records, Reports, Tests, and Inspections

2.5.2.1 Introduction

This chapter evaluates the U.S. Department of Energy (DOE) Safety Analysis Report (SAR) Section 5.2 (DOE, 2008ab, 2009av) description of the program to be used for maintaining records of the receipt, handling, and disposition of radioactive waste; maintaining construction records; retaining records in a manner that ensures their usability for future generations; reporting deficiencies to NRC; performing tests or allowing NRC to perform tests; and allowing NRC to inspect the premises of the geologic repository operations area (GROA) at the Yucca Mountain site. DOE provided further information on these topics in its response to an NRC staff request for additional information (DOE, 2009be).

2.5.2.2 Evaluation Criteria

10 CFR 63.21(c)(23) requires that the SAR include a description of the program to be used to maintain the records described in 10 CFR 63.71 and 63.72. 10 CFR 63.73 provides requirements for reporting deficiencies to the NRC, 10 CFR 63.74 provides requirements for performing tests or allowing the NRC to perform tests, and 10 CFR 63.75 provides requirements for allowing NRC to inspect the GROA at the Yucca Mountain site and adjacent areas to which DOE has rights of access.

The NRC staff evaluated the DOE's program for maintaining records using guidance in the Yucca Mountain Review Plan (YMRP) Section 2.5.2.3 (NRC, 2003aa). The YMRP Acceptance Criterion relating to records, reports, tests, and inspections states: "DOE will maintain adequate records and reports that may be required by conditions of the license or rules, regulations, and orders of the Commission."

2.5.2.3 Technical Evaluation

SAR Sections 5.2.1 and 5.2.2 contain information on records and reports. SAR Sections 5.2.1.3 and 5.2.1.4 state that procedures for record maintenance and storage will incorporate guidance from NRC Regulatory Issue Summary 2000-18 (NRC, 2000ad) and American National Standards Institute/American Society of Mechanical Engineers (ANSI/ASME) Standard NQA–1–1983 (American Society of Mechanical Engineers, 1983aa), respectively. DOE described the recordkeeping and reporting programs for receipt, handling, and disposition of radioactive waste to provide a complete history of the movement of the waste from the shipper through all phases of storage and disposal. For illustrative purposes (DOE, 2009be), DOE identified regulations addressing records and reporting (e.g., 10 CFR Part 21) in SAR Tables 5.2-1 and 5.2-2.

SAR Section 5.2.1.2 includes the construction records that DOE stated that it will create. SAR Sections 5.2.1.3 and 5.2.1.4 state that procedures for record maintenance and storage will incorporate guidance from NRC Regulatory Issue Summary 2000-18 (NRC, 2000ad) and ANSI/ASME Standard NQA–1–1983 (American Society of Mechanical Engineers, 1983aa), respectively. DOE described a program to maintain construction records of the GROA at the Yucca Mountain site in a manner to ensure their usability for future generations.

NRC staff reviewed DOE's program for reporting deficiencies to NRC. DOE stated that methods will be in place to evaluate and report deviations and failures to comply, as well as to identify defects and failures to comply, that are associated with substantial safety hazards at the GROA. Methods will also be in place to address reporting specific events and conditions. The discussion in SAR Section 5.2.2 and in DOE (2009be) describes reporting for both emergency and nonemergency events and conditions.

In SAR Section 5.2.3, DOE described a program for DOE to perform, or permit NRC to perform, tests that NRC considers appropriate or necessary. DOE's testing program will include implementing the Performance Confirmation Program. The NRC staff evaluates the DOE's description of its Performance Confirmation Program in TER Section 2.4.

In SAR Section 5.2.4, DOE stated that it will provide immediate and unfettered access for NRC personnel to the GROA and adjacent areas and access to DOE records, upon reasonable notice. DOE also stated that it will provide office space for the exclusive use of NRC inspection personnel.

On the basis of its review, the NRC staff notes that DOE's description of the program to be used for maintaining records of the receipt, handling, and disposition of radioactive waste; maintaining construction records; retaining records; reporting deficiencies to NRC; performing tests or allowing NRC to perform tests; and allowing NRC to inspect the premises of the geologic repository operations area (GROA) is reasonable because (i) the recordkeeping and reporting programs for receipt, handling, and disposition of radioactive waste provides a complete history of the movement of the waste from the shipper through all phases of storage and disposal; (ii) the procedures for record retention, maintenance, and storage are consistent with guidance from NRC Regulatory Issue Summary 2000-18 (NRC, 2000ad) and ANSI/ASME Standard NQA–1–1983 (American Society of Mechanical Engineers, 1983aa); (iii) DOE stated methods will be in place to address reporting specific events and conditions prior to the receipt of radioactive waste being received at the site; (iv) the testing program permits NRC to perform tests that NRC considers appropriate or necessary; and (v) DOE stated it will provide immediate and unfettered access for NRC personnel to the GROA and adjacent areas and access to DOE records, upon reasonable notice.

2.5.2.4   NRC Staff Conclusions

The NRC staff notes that DOE's descriptions of the programs for records, reports, tests, and inspections are reasonable. The information DOE provided with respect to the programs for records, reports, tests, and inspections is consistent with the guidance in the YMRP.

2.5.2.5   References

American Society of Mechanical Engineers. 1983aa. ANSI/ASME NQA–1–1983, "Quality Assurance Program Requirements for Nuclear Facilities." New York City, New York: American Society of Mechanical Engineers.

DOE. 2009av. DOE/RW–0573, "Yucca Mountain Repository License Application." Rev. 1. ML090700817. Las Vegas, Nevada: DOE, Office of Civilian Radioactive Waste Management.

DOE. 2009be. "Yucca Mountain—Response to Request for Additional Information Regarding License Application (Safety Analysis Report Section 5.2), Safety Evaluation Report Vol. 4, Chapter 2.5.2, Set 1." Letter (January 16) J.R. Williams to B.J. Benney (NRC). ML090210487. Washington, DC: DOE, Office of Technical Management.

DOE. 2008ab. DOE/RW–0573, "Yucca Mountain Repository License Application." Rev. 0. ML081560400. Las Vegas, Nevada: DOE, Office of Civilian Radioactive Waste Management.

NRC. 2003aa. NUREG–1804, "Yucca Mountain Review Plan—Final Report." Rev. 2. ML032030389. Washington, DC: NRC.

NRC. 2000ad. "NRC Regulatory Issue Summary 2000-18 Guidance on Managing Quality Assurance Records in Electronic Media." ML003739359. Washington, DC: NRC.

CHAPTER 5

2.5.3.1 U.S. Department of Energy Organizational Structure As It Pertains to Construction and Operation of Geologic Repository Operations Area

2.5.3.1.1 Introduction

This chapter evaluates the U.S. Department of Energy (DOE) organizational structure, as it pertains to construction and operation of the geologic repository operations area (GROA), provided in the DOE Safety Analysis Report (SAR) Section 5.3.1 (DOE, 2008ab, 2009av), as supplemented by its response to an U.S. Nuclear Regulatory Commission (NRC) request for additional information (DOE, 2009az). SAR Section 5.3.1 described any delegations of authority and assignments of responsibilities.

2.5.3.1.2 Evaluation Criteria

10 CFR 63.21(c)(22)(i) requires the SAR to include the organizational structure pertaining to construction and operation of the GROA at the Yucca Mountain site, and a description of any delegations of authority and assignments of responsibilities.

The NRC staff evaluated the organizational structure using guidance in the Yucca Mountain Review Plan (YMRP) Section 2.5.3.1 (NRC, 2003aa). The YMRP acceptance criteria are (i) DOE assignments of responsibility are adequately defined and (ii) an adequate procedure for delegation of authority situations is in place.

2.5.3.1.3 Technical Evaluation

In SAR Section 5.3.1 and in DOE (2009az), DOE described the organizational structure anticipated at the time of repository construction and operations for the GROA at the Yucca Mountain site, including a description of a procedure for delegation of authority (SAR Section 5.3.1.4). DOE described the responsibilities of the director, management functions and responsibilities, reporting relationships, and principal lines of communication. DOE also provided descriptions of the Executive Advisory Board responsible for advising the director regarding executive-level matters, each of the eight key staff assigned responsibility for safety and operations at the site, and the Onsite Safety Committee responsible for advising the site operations manager regarding operations matters. DOE (2009az) provided a revised organizational structure (revised SAR Figure 5.3-1) that DOE stated will allow the nuclear criticality safety program to be administratively independent of operations, to function organizationally, and to have responsibilities assigned in a manner consistent with those of other safety programs, such as Radiation Protection. In SAR Section 5.3.1, DOE stated that it will update, as necessary, specific contact information that includes the address of the office of record for each entity in the organization who holds a key onsite or offsite position, a point of contact, a telephone number, a fax number, and an e-mail address.

On the basis of the information in SAR Section 5.3.1 and in DOE (2009az), the NRC staff notes that DOE provided a reasonable organizational structure pertaining to construction and operation of the GROA because: (i) DOE described the delineation of responsibility and decision-making authority during construction and operation of the GROA for the management, staff, and affected organizations and (ii) procedures are in place for delegation of authority for positions having responsibility to act in routine or emergency situations, assuring that an identified party will always have responsibility and sufficient authority to act, along with appropriate qualifications.

2.5.3.1.4　　　　NRC Staff Conclusions

The NRC staff notes that DOE has provided a reasonable description of the organizational structure for the construction and operation of the GROA, including the description of any delegations of authority and assignments of responsibilities. The DOE's organizational structure is consistent with the guidance in the YMRP.

2.5.3.1.5　　　　References

DOE. 2009av. DOE/RW–0573, "Yucca Mountain Repository License Application." Rev. 1. ML090700817. Las Vegas, Nevada: DOE, Office of Civilian Radioactive Waste Management.

DOE. 2009az. "Yucca Mountain—Response to Request for Additional Information Regarding License Application (Safety Analysis Report Sections 5.3 and 1.14.1), Safety Evaluation Report Vol. 4, Chapter 2.5.3.2, Set 1." Letter (February 10) J.R. Williams to B.J. Benney (NRC). ML090420250. Washington, DC: DOE, Office of Technical Management.

DOE. 2008ab. DOE/RW–0573, "Yucca Mountain Repository License Application." Rev. 0. ML081560400. Las Vegas, Nevada: DOE, Office of Civilian Radioactive Waste Management.

NRC. 2003aa. NUREG–1804, "Yucca Mountain Review Plan—Final Report." Rev. 2. ML032030389. Washington, DC: NRC.

CHAPTER 6

2.5.3.2 Key Positions Assigned Responsibility for Safety and Operations of Geologic Repository Operations Area

2.5.3.2.1 Introduction

This chapter evaluates the description in the U.S. Department of Energy (DOE) Safety Analysis Report (SAR) Section 5.3.2 (DOE, 2008ab, 2009av) for key positions assigned responsibility for safety and operations at the repository site. DOE provided more information on this topic in its response to the NRC staff's request for additional information (DOE, 2009az).

2.5.3.2.2 Evaluation Criteria

10 CFR 63.21(c)(22)(ii) requires that the SAR include identification of the key positions that are assigned responsibility for safety at, and operation of, the geologic repository operations area (GROA). The NRC staff evaluated DOE's plans for key positions for safety and operations at the GROA, using guidance in the Yucca Mountain Review Plan (YMRP) Section 2.5.3.2 (NRC, 2003aa). The acceptance criterion in the YMRP for the key positions states

"The description of the key positions for safety at the GROA are adequate

— DOE provides an adequate description of each key position that includes the minimum skills and experience necessary to hold the position; and

— Qualified alternates to act in the absence of assigned individuals are identified based on minimum skills and experience necessary to hold the each key position."

As indicated in the YMRP (p. 2.5-52), DOE is not expected to have identified specific individuals to fill key positions for a construction authorization. Therefore, NRC staff review of this information on key positions may await an application to receive and possess radioactive material at the repository.

2.5.3.2.3 Technical Evaluation

In SAR Section 5.3.2 and in DOE (2009az), DOE described key positions and the responsibilities and qualifications of those holding the positions; in SAR Section 5.3.1.4, DOE described its plan for identifying qualified alternates who would act in the absence of individuals assigned to key GROA positions. DOE provided functional titles and qualifications for the following key positions: site operations manager, quality assurance manager, engineering and construction manager, licensing manager, postclosure performance and confirmation manager, site protection manager, radiation protection manager, and operations manager. Additionally, DOE provided more than 10 functional titles and qualifications for key positions under the engineering and construction manager or the site protection manager (e.g., emergency preparedness manager) or the operations manager (e.g., criticality safety manager, training manager). In DOE (2009az), DOE provided a revised description for the radiation protection manager that incorporated the responsibilities and qualifications for the criticality safety manager.

On the basis of its review, the NRC staff notes that the DOE description of the key positions, their responsibilities, and qualifications is reasonable because the DOE description (i) identified more than 20 key positions responsible for safety and operations, including construction activities that are expected to take place during GROA operation; (ii) provided the minimum skills and experience levels for personnel filling the key positions and their alternates; and (iii) provided a plan for how qualified alternates will be identified to act in the absence of DOE staff assigned to key positions. DOE explained that the minimum skills and experience necessary to hold each key position will be considered in determining a qualified alternate. The NRC staff also notes that the minimum skills and experience levels DOE described are applicable and consistent with the nature and responsibilities of the key positions.

2.5.3.2.4 NRC Staff Conclusions

The NRC staff notes that DOE reasonably described the key positions assigned responsibility for GROA safety and operations and the qualifications of the persons occupying these positions. The DOE's description of the key positions and qualifications (i.e., minimum skills and experience) is consistent with the guidance in the YMRP, which, as discussed in TER Section 2.5.3.2.2, recognizes that DOE is not expected to have identified specific individuals to fill key positions for a construction authorization.

2.5.3.2.5 References

DOE. 2008ab. DOE/RW–0573, "Yucca Mountain Repository License Application." Rev. 0. ML081560400. Las Vegas, Nevada: DOE, Office of Civilian Radioactive Waste Management.

DOE. 2009av. DOE/RW–0573, "Yucca Mountain Repository License Application." Rev. 1. ML090700817. Las Vegas, Nevada: DOE, Office of Civilian Radioactive Waste Management.

DOE. 2009az. "Yucca Mountain—Response to Request for Additional Information Regarding License Application (Safety Analysis Report Sections 5.3 and 1.14.1), Safety Evaluation Report Vol. 4, Chapter 2.5.3.2." Letter (February 10) J.R. Williams to B. Benney (NRC). Enclosures (1). ML090420250. Las Vegas, Nevada: DOE, Office of Civilian Radioactive Waste Management.

NRC. 2003aa. NUREG–1804, "Yucca Mountain Review Plan—Final Report." Rev. 2. ML032030389. Washington, DC: NRC.

CHAPTER 7

2.5.3.3 Personnel Qualifications and Training Requirements

2.5.3.3.1 Introduction

This chapter evaluates the personnel qualifications and training program requirements in the U.S. Department of Energy (DOE) Safety Analysis Report (SAR) Section 5.3 (DOE, 2008ab, 2009av).

2.5.3.3.2 Evaluation Criteria

10 CFR 63.21(c)(22)(iii) requires that the SAR include the personnel qualifications and training requirements for activities at the geologic repository operations area (GROA). DOE's personnel qualifications and training requirements must address the general requirements, the training and certification program, and the physical requirements as specified in 10 CFR Part 63, Subpart H, Training and Certification of Personnel.

The NRC staff evaluated DOE's personnel qualifications and training requirements using the guidance in the Yucca Mountain Review Plan (YMRP) Section 2.5.3.3 (NRC, 2003aa). The acceptance criteria for personnel qualifications and training are as follows:

- Adequate standards are used for selection, training, and certification of personnel.

- Programs for general training, proficiency testing, and certification of GROA personnel are acceptable.

- An acceptable preoperational and operational radioactive materials training program is provided.

- Operation of equipment and controls identified as important to safety is limited to trained and certified personnel or is under the direct visual supervision of an individual with training and certification in their operation.

- An acceptable operator and supervisor requalification program for structures, systems and components (SSCs) important to safety is provided.

- Physical condition and general health of personnel certified for the operation of equipment and controls important to safety are such that operational errors that could endanger other in-plant personnel or the public health and safety will not occur.

- Methods for selecting, training, and qualifying security guards are acceptable.

- Methods used to evaluate operator testing procedures are acceptable.

- Qualifications of personnel are adequate.

As indicated in the YMRP (p. 2.5-53), DOE is not expected to have an NRC-approved personnel qualifications and training program in place for a construction authorization.

2.5.3.3.3   Technical Evaluation

In SAR Section 5.3 (DOE, 2008ab, 2009av), DOE described its training program for the operational phase of the repository, as well as preoperational, functional, and initial startup testing. In general, the training program applies to personnel engaged in operations, maintenance, testing, or other activities of the repository structures, systems, and components (SSCs) that are important to safety (ITS) or important to waste isolation (ITWI). In particular, DOE noted the following, with respect to its personnel qualifications and training program:

- Training and certification programs will be implemented in time to provide the training and certification of personnel before work activities are performed.

- Formal, documented training programs will be established for personnel assigned to the repository, including methods for verifying training effectiveness (e.g., written tests and actual demonstration of skills).

- Training programs will include general employee training provided within 30 days of reporting to work, consisting of safety preparedness for all safety disciplines, training on emergency preparedness, training practices in keeping doses as low as is reasonably achievable (ALARA), and specific training on in-depth technical subjects.

- General employee training will be given to those employees at the repository who are permitted inside the GROA without a full-time escort.

- A needs and job analysis will be performed, and tasks will be identified to develop training for personnel working on tasks that are ITS or ITWI.

- Training programs will be established for job positions commensurate with the criticality potential or radiation safety responsibilities.

- Qualification and training for personnel will be established and implemented in DOE procedures.

- Personnel performing activities relied upon for safety and waste isolation will be evaluated every 2 years.

- Personnel, designated as operators of equipment or controls that are ITS or ITWI, will have a physical examination every 2 years.

The NRC staff reviewed SAR Section 4.3 and notes that DOE's description of the personnel qualifications and training program is reasonable because DOE stated that its personnel qualifications and training program will include (i) standards for training and certification of personnel, (ii) general training, (iii) preoperational and operational training, (iv) training for operators of equipment and controls ITS or ITWI, (v) requalification program, (vi) physical condition and health of personnel, (vii) methods for evaluating training effectiveness, and (viii) qualifications that are set for personnel based on safety responsibilities. Additionally, as described in NRC Section 1.3.3.2.4 (2010aa), DOE's methods for selecting, training, and qualifying members of the security organization were reviewed and determined to be reasonable because they were consistent with the YMRP regarding training, equipping, qualifying, and requalifying members of the security organization.

2.5.3.3.4   NRC Staff Conclusions

The NRC staff notes that DOE provided reasonable information on the personnel qualifications and training program that DOE stated will be implemented before DOE receives, possesses, stores, or disposes high-level radioactive waste. The DOE's information regarding the personnel qualifications and training program is consistent with the guidance in the YMRP, which recognizes, as discussed in TER Section 2.5.3.3.2, that DOE is not expected to have an NRC-approved personnel qualifications and training program in place for a construction authorization.

2.5.3.3.5   References

DOE. 2009av. DOE/RW–0573, "Yucca Mountain Repository License Application." Rev. 1. ML090700817. Las Vegas, Nevada: DOE, Office of Civilian Radioactive Waste Management.

DOE. 2008ab. DOE/RW–0573, "Yucca Mountain Repository License Application." Rev. 0. ML081560400. Las Vegas, Nevada: DOE, Office of Civilian Radioactive Waste Management.

NRC. 2010aa. NUREG–1949, "Safety Evaluation Report Related to Disposal of High-Level Radioactive Wastes in a Geologic Repository at Yucca Mountain, Nevada. Vol. 1: General Information." Washington, DC: NRC.

NRC. 2003aa. NUREG–1804, "Yucca Mountain Review Plan—Final Report." Rev. 2. ML032030389. Washington, DC: NRC.

CHAPTER 8

2.5.5 Plans for Startup Activities and Testing

2.5.5.1 Introduction

This chapter evaluates the description of plans for startup activities and testing provided in the U.S. Department of Energy (DOE) Safety Analysis Report (SAR) Section 5.5 (DOE, 2008ab, 2009av). Additional information was provided in DOE's response to an NRC staff request for additional information regarding human factors considerations (DOE, 2009go).

2.5.5.2 Evaluation Criteria

10 CFR 63.21(c)(22)(iv) requires that the SAR include plans for startup activities and startup testing at the geologic repository operations area (GROA).

The NRC staff evaluated DOE's plans for startup activities and testing using the guidance in the Yucca Mountain Review Plan (YMRP) Section 2.5.5 (NRC, 2003aa). The acceptance criteria state the following:

- Systems used to develop, review, and approve individual pre-startup test procedures are acceptable.

- Summaries of pre-startup test programs and objectives are adequate.

- Design performance information is adequately incorporated in pre-startup testing plans.

- Format and content of test procedures are acceptable.

- Test descriptions are acceptable.

- Test programs are consistent with applicable regulatory guidance.

- Adequate use is made of experience from similar facilities.

- Initial operating procedures will protect workers and the public.

- Schedules for each phase of the testing program are acceptable.

- Structures, systems, and components important to safety whose functional adequacy has not been demonstrated by prior use or otherwise validated are tested and evaluated before the receipt of radioactive waste.

- Plans for initial startup of geologic repository operations area structures, systems, and components important to safety and integrated operation of the GROA are acceptable.

- Overall GROA safety is adequately supported by facility startup and testing plans.

As indicated in the YMRP (p. 2.5-66), DOE is not expected to have prepared plans for startup activities and testing for a construction authorization.

2.5.5.3   Technical Evaluation

SAR Section 5.5 described

- The testing programs

- Use of experience from similar activities

- Test procedure development, approval by authorized personnel, and evaluation of test results

- Format and content of test procedures

- Component testing

- Systems functional testing

- Cold integrated systems testing

- Operational readiness review

- Protection of workers and the public

- Hot testing, which includes initial startup operations

- The schedules for startup activities and testing

- Testing and evaluating functional adequacy of new or untested structures, systems, and components (SSCs)

DOE (2009go) clarified how it will incorporate human factors consideration into its assessment of testing results and operational lessons learned from similar facilities:

- With respect to the description of the systems used to develop, review, and approve individual pre-startup test procedures, DOE stated that (i) startup testing will be conducted using written, reviewed, and approved procedures; (ii) test procedures for the first phase of repository operations will be prepared, approved, and implemented prior to submittal of the updated license application for a license to receive and possess spent nuclear fuel and high-level radioactive waste (HLW); (iii) personnel preparing the test procedures will be qualified in accordance with the Testing Program Plan; and (iv) individuals leading these efforts will be experienced and qualified in the areas for which they are preparing procedures or performing tests.

- The SAR described the various types of tests and objectives for the tests to be performed for each GROA SSC important to safety. DOE stated that test procedures will include the type of tests to be performed, expected response, acceptable margin of difference from the expected response, method of test validation, and appropriateness of proposed corrective action for unexpected or unacceptable test results.

- The SAR indicated that design information and data from preconstruction performance tests or evaluations will be considered in development of the startup testing procedures. Test results will be evaluated to determine whether the design safety basis has been met.

- The SAR stated that a testing program plan will be developed that describes a consistent test procedure format. The SAR specified the content requirements for procedures, including purpose and role of the test, prerequisites, types, objectives, test acceptance criteria, and methods for determining corrective action.

- The SAR stated that test procedures will address objectives, design specifications and other requirements, prerequisites, objectives, test methods, and test acceptance criteria.

- DOE stated that its startup testing plans and procedures will be consistent with regulatory guidance where the guidance is applicable to the repository structures, systems, and components (SSCs) and the repository operations. DOE also stated it would provide justification for positions that deviate from the regulatory guidance as plans and procedures are developed.

- The SAR and DOE (2009go) described a program for managing operating experience and good work practices from DOE sites, NRC-licensed facilities, and other similar facilities. DOE stated that the results will be evaluated and incorporated into test procedures.

- DOE stated that test procedures will address prerequisites and precautionary measures needed to protect workers and the public during startup activities. To the extent practicable, each phase of startup testing will be conducted before a radiation source is placed in the area.

- The SAR described a schedule, over distinct time intervals, corresponding to the phased construction of facilities. Prior to receiving waste at each facility, the startup and testing program will test the capability for handling and processing waste and for limiting the release of radioactive materials. Detailed procedures will be developed in a phased manner to support the initial operation of the facility. DOE stated that detailed procedures will be in place prior to the receipt of waste.

- DOE stated that, prior to receipt of waste, new and untested configurations of SSCs would be tested and evaluated. Test procedure development will account for the first-of-a-kind applications. Dry runs would be considered to ensure functional adequacy of such SSCs.

- DOE test plans include cold integrated systems testing and hot testing of each operation involving radioactive waste streams. DOE stated that an operational readiness review will be performed at each facility and phase of the repository prior to initial startup operations. Hot testing and initial startup operations will be conducted to evaluate the ability to receive, package, and emplace waste in a manner that protects the workers and the public.

- The SAR included a comprehensive and integrated approach to initial startup activities and testing that will assess whether SSCs important to safety and waste isolation have been properly constructed and installed and will fulfill their operational and safety functions in accordance with the DOE design basis.

The NRC staff reviewed SAR Section 5.5 and notes that DOE's description of the plans for startup activities and testing is reasonable because DOE described (i) how pre-startup test procedures would be developed, reviewed, and approved; (ii) how the test programs would be conducted, including the test objectives and schedules; (iii) how design performance is incorporated into the test plans; (iv) how test procedures will address objectives, design specifications and other requirements, prerequisites, objectives, and test methods; (v) how it would provide justification for positions that deviate from regulatory guidance as plans and procedures are developed; (vi) how experience from similar facilities is used to develop test procedures; (vii) how test procedures will address prerequisites and precautionary measures needed to protect workers and the public; (viii) how detailed procedures will be developed in a phased manner to support the initial operation of the facility; and (ix) how initial startup activities and testing will assess whether SSCs important to safety and waste isolation have been properly constructed and installed, consistent with the DOE design basis.

2.5.5.4 NRC Staff Conclusions

The NRC staff notes that DOE has reasonably described the plans for startup activities and testing that DOE stated will be implemented before DOE receives, processes, stores, or disposes high-level radioactive waste. The DOE's description of its plans for startup activities and testing is consistent with the guidance in the YMRP, which, as discussed in TER Section 2.5.5.2, recognizes that DOE is not expected to have prepared plans for startup activities and testing for a construction authorization.

2.5.5.5 References

DOE. 2009av. DOE/RW–0573, "Yucca Mountain Repository License Application." Rev. 1. ML090700817. Las Vegas, Nevada: DOE, Office of Civilian Radioactive Waste Management.

DOE. 2009go. "Yucca Mountain—Response to Request for Additional Information Regarding License Application (Safety Analysis Report Section 5.5), Safety Evaluation Report Vol. 4, Chapter 2.5.5, Set 1, and Yucca Mountain—Response to Request for Additional Information Regarding License Application (Safety Analysis Report Section 5.6), Safety Evaluation Report Vol. 4, Chapter 2.5.6, Set 1." Letter (July 9) J.R. Williams to F. Jacobs (NRC). ML091900678. Washington, DC: DOE, Office of Technical Management.

DOE. 2008ab. DOE/RW–0573, "Yucca Mountain Repository License Application." Rev. 0. ML081560400. Las Vegas, Nevada: DOE, Office of Civilian Radioactive Waste Management.

NRC. 2003aa. NUREG–1804, "Yucca Mountain Review Plan—Final Report." Rev. 2. ML032030389. Washington, DC: NRC.

NRC. 1989aa. Regulatory Guide 3.48, "Standard Format and Content for the Safety Analysis Report for an Independent Spent Fuel Storage Installation or Monitored Retrievable Storage Installation (Dry Storage)." Rev. 1. Washington, DC: NRC.

NRC. 1978aa. Regulatory Guide 1.68, "Initial Test Programs for Water-Cooled Nuclear Power Plants." Rev. 2. Washington, DC: NRC.

CHAPTER 9

2.5.6 Plans for Conduct of Normal Activities, Including Maintenance, Surveillance, and Periodic Testing

2.5.6.1 Introduction

This chapter of the Technical Evaluation Report (TER) evaluates the description of plans provided in the U.S. Department of Energy (DOE) Safety Analysis Report (SAR) Section 5.6 (DOE, 2008ab, 2009av) for conduct of normal activities, including those activities necessary to maintain and verify proper operations. DOE provided more information on this topic in its response to the NRC staff's request for additional information (DOE, 2009go).

2.5.6.2 Evaluation Criteria

10 CFR 63.21(c)(22)(v) requires that the SAR include plans for conduct of normal activities, including maintenance, surveillance, and periodic testing of structures, systems, and components (SSCs) of the geologic repository operations area (GROA).

The NRC staff evaluated DOE's plans for conduct of normal activities, using the guidance in the Yucca Mountain Review Plan (YMRP) Section 2.5.6 (NRC, 2003aa). The relevant acceptance criteria follow:

- Plans for normal operations of structures, systems, and components (SSCs) of the GROA that are important to safety are acceptable.

- Plans and procedures for maintenance of SSCs of the GROA that are important to safety are acceptable.

- Plans and procedures for surveillance of SSCs of the GROA that are important to safety are acceptable.

- Plans and procedures for routine periodic testing of SSCs of the GROA that are important to safety are acceptable.

As indicated in the YMRP, DOE is not expected to have plans in place for normal activities for a construction authorization. DOE would develop and implement the plans prior to receipt and possession of waste (YMRP p. 2.5-74).

2.5.6.3 Technical Evaluation

SAR Section 5.6 described the plan and procedure development, testing, and approval by authorized personnel; management systems for operation of the repository, including administrative and procedural safety controls; and the specific types of plans and procedures to be developed for normal operations, maintenance, and periodic surveillance testing. DOE (2009go) stated that it would revise the SAR to include human factors engineering as a competency of the independent procedure review personnel. DOE stated that it operates many facilities with features similar to the repository surface handling facilities and also operates the Waste Isolation Pilot Plant (SAR p. 5.6-6). DOE identified experience from these similar DOE facilities that it would use in developing plans and procedures for conduct of normal activities at the GROA.

With respect to its plans for normal activities, DOE stated the following:

- Controls will be applied to develop plans and procedures, including content and format, review, approval, and change controls

- Plans and procedures will address the full range of normal activities for SSCs important to safety, including maintenance, surveillance, and routine periodic testing

- Plans and procedures will be developed in a phased manner to support the operation of each handling facility as the construction of each facility is completed, prior to the receipt of waste

- Plans and procedures for operations, maintenance, surveillance, and periodic testing of SSCs and processes, including procedural safety controls, will be written, tested, and approved by the waste handling manager prior to receipt of waste

- Plans and procedures will be reviewed by safety, health, environmental, and quality assurance personnel independent of the operating organization with appropriate technical qualifications, including human factors engineering

- Operating experience and lessons learned from other facilities, operated by both DOE and other entities, will be incorporated into plans and procedures as appropriate

- Operations personnel will be appropriately trained and qualified, and maintenance assignments will consider qualification and training of personnel

The NRC staff reviewed SAR Section 5.6 and notes that DOE's description of the plans for normal activities, including maintenance, surveillance, and periodic testing, is reasonable because DOE described (i) plans and procedures for normal operations and that the plans and procedures will address the full range of normal activities for SSCs important to safety and (ii) the plans and procedures for operations, maintenance, surveillance, and periodic testing of SSCs and processes, including procedural safety controls.

2.5.6.4 NRC Staff Conclusions

The NRC staff notes that DOE reasonably described plans for conduct of normal activities, including maintenance, surveillance, and periodic testing that DOE stated will be implemented before receipt, possession, storing, or disposal of high-level radioactive waste. The DOE's description of its plans for normal activities, including maintenance, surveillance, and periodic testing, is consistent with the guidance in the YMRP, which as discussed in TER Section 2.5.6.2, recognizes that DOE is not expected to have plans in place for normal activities for a construction authorization.

2.5.6.5 References

DOE. 2009av. DOE/RW–0573, "Yucca Mountain Repository License Application." Rev. 1. ML090700817. Las Vegas, Nevada: DOE, Office of Civilian Radioactive Waste Management.

DOE. 2009go. "Yucca Mountain—Response to Request for Additional Information Regarding License Application (Safety Analysis Report Section 5.5), Safety Evaluation Report Vol. 4, Chapter 2.5.5, Set 1, and Yucca Mountain—Response to Request for Additional Information Regarding License Application (Safety Analysis Report Section 5.6), Safety Evaluation Report Vol. 4, Chapter 2.5.6, Set 1." Letter (July 9) J.R. Williams to F. Jacobs (NRC). ML091900678. Washington, DC: DOE, Office of Technical Management.

DOE. 2008ab. DOE/RW–0573, "Yucca Mountain Repository License Application." Rev. 0. ML081560400. Las Vegas, Nevada: DOE, Office of Civilian Radioactive Waste Management.

NRC. 2003aa. NUREG–1804, "Yucca Mountain Review Plan—Final Report." Rev. 2. ML032030389. Washington, DC: NRC.

CHAPTER 10

2.5.7 Emergency Planning

2.5.7.1   Introduction

This chapter of the Technical Evaluation Report (TER) evaluates the description provided in the U.S. Department of Energy (DOE) Safety Analysis Report (SAR) Section 5.7 (DOE, 2008ab, 2009av) of plans for responding to, and recovering from, radiological emergencies that may occur at a geologic repository operations area (GROA) for disposal of high-level waste at any time before permanent closure. DOE provided more information on this topic in its response to an NRC staff request for additional information (DOE, 2008ai).

Some topics evaluated in this chapter are also addressed, in part, in other TER chapters. Training, as related to the training and certification of personnel, is evaluated in Chapter 2.5.3. The types of accidents that could lead to radiological accidents are evaluated in the TER Preclosure Volume (Chapter 2.1.1.4). Although the training and the types of accident topics are evaluated in other chapters of the TER, the evaluation in this chapter does not rely on those chapters' evaluations.

2.5.7.2   Evaluation Criteria

10 CFR 63.21(c)(21) and 10 CFR Part 63, Subpart I, Emergency Planning Criteria, contain the regulatory requirements for emergency planning. 10 CFR 63.21(c)(21) requires that the SAR include a description of the plan for responding to, and recovering from, radiological emergencies that may occur any time before permanent closure and decontamination or decontamination and dismantlement of surface facilities, as required by 10 CFR 63.161. 10 CFR Part 63, Subpart I contains 10 CFR 63.161, which requires that DOE (i) develop and be prepared to implement a plan to cope with radiological accidents that may occur at the GROA at any time before permanent closure and decontamination or decontamination and dismantlement of surface facilities and (ii) base the emergency plan on the criteria of 10 CFR 72.32(b).

NRC staff used guidance in Yucca Mountain Review Plan (YMRP) Section 2.5.7 (NRC, 2003aa). YMRP Section 2.5.7 references NUREG–1567, Standard Review Plan for Spent Fuel Storage Facilities (NRC, 2000ab), as additional guidance. NUREG–1567, Section 10.4.5, Emergency Planning, was revised by Spent Fuel Project Office Interim Staff Guidance (ISG)–16, Emergency Planning (NRC, 2000aa). The YMRP acceptance criterion is that an adequate emergency plan for responding to potential radiological materials and other accidents at the GROA is provided. The YMRP (p. 2.5-88) also states that the NRC staff review should be conducted using information that is reasonably available.

2.5.7.3   Technical Evaluation

In SAR Section 5.7, DOE provided a description of its Emergency Plan for responding to, and recovering from, radiological emergencies that may occur during operations at the repository. DOE stated that the information provided was as complete as possible, in light of information that was reasonably available at this time. DOE stated it would provide an Emergency Plan to NRC no later than 6 months prior to the submittal of the updated application for a license to receive and possess high-level radioactive waste.

DOE's description of its Emergency Plan is divided into 16 specific topics.

1.   Facility Description

In SAR Section 5.7.2, DOE provided information that it stated would be included in the Emergency Plan to describe the GROA and the surrounding area. DOE stated that the Emergency Plan will include detailed maps of the GROA and the site, and DOE expects to provide enlarged duplicates of the drawings suitable for use as wall maps to NRC with the updated Emergency Plan. DOE stated the Emergency Plan will concisely describe site features affecting emergency response, including communications and assessment centers, assembly and relocation areas, and emergency equipment storage areas. DOE addressed the description of the area near the site and stated that the Emergency Plan will describe the GROA and the surrounding area and will include a general map.

2.   Types of Accidents

In SAR Section 5.7.3.1, DOE stated that the Emergency Plan will discuss each type of potential accident that could result in the release of radioactive material, as identified in the list of possible internal and external events presented in SAR Section 1.7. DOE will use the event sequence analysis, which is part of the preclosure safety analysis, to identify each type of potential radioactive materials accident. DOE (2008ai) stated it would update the SAR to state that the Emergency Plan will also include descriptions of any non-radiological, hazardous material release events that could impact emergency response efforts.

3.   Classification of Accidents

In SAR Section 5.7.3.2.1, DOE stated that the emergency action levels (EALs) will be specific, predetermined, observable criteria that will be used to determine the emergency classification and the initial protective actions required for those emergencies which might be most likely to occur at the repository. In addition, DOE stated that the Emergency Plan will contain EALs, providing initiating conditions, accident mechanisms, postulated equipment or system failures, event indicators, and contributing events. Finally, DOE described Category 1 and Category 2 events that could lead to the initiation of an alert and site area emergency.

4.   Detection of Accidents

In SAR Section 5.7.4.1, DOE stated that the Emergency Plan will provide a description of the means for detecting initiating events and accident conditions that apply to each identified accident. The Emergency Plan will also describe the rationale for the locations and types of devices used to detect accidents as identified in the list of possible internal and external events presented in SAR Section 1.7. In SAR Section 5.7.4.2, DOE stated that the Emergency Plan will include operating procedures that will identify the means of detection of the event sequences that lead to declaration of an alert or site area emergency. DOE explained that this may include, but is not limited to, visual observations, radiation monitors, smoke or heat detectors, and/or process alarms.

5.   Mitigation of Consequences

In SAR Section 5.7.5.1, DOE stated that the Emergency Plan will describe the means to mitigate consequences of each type of accident, as presented in SAR Section 1.7. DOE stated that these descriptions will address

- Equipment and design features relied on to mitigate emergencies

- General actions that repository personnel may take to mitigate emergencies

- Protective actions to be taken to protect the health and safety of workers and the public

- Arrangements for first aid, medical, and hospital services and underground rescue

- Facilities available to support mitigation efforts

- Types and locations of response and communication equipment available to support mitigation efforts

- Processes for periodically inventorying, testing, and maintaining emergency equipment, including mitigation equipment

DOE (2008ai) additionally stated it would update the SAR so that the Emergency Plan would include (i) a description of the criteria for the shutdown of systems or facility(s); (ii) the steps to be taken to ensure a safe, orderly shutdown; and (iii) the approximate time required for a shutdown based on the type of emergency.

6.   Assessment of Releases

In SAR Section 5.7.6.1, DOE stated that the Emergency Plan will describe radiological sampling and monitoring methods, instrumentation, equipment, and procedures to be used to assess the extent of radiological releases. The Emergency Plan will also identify organizational positions for which training and qualification for assessment of radioactive releases is required.

7.   Roles and Responsibilities for Repository Personnel During an Emergency

In SAR Section 5.7.7.1, DOE stated that the Emergency Plan will identify the personnel responsible for ensuring that offsite notifications are performed promptly. The Emergency Plan will also identify the means to ensure that the communication chain for notifying and mobilizing emergency response personnel is maintained during normal and off-normal working hours (e.g., nights, weekends, and holidays). Additionally, DOE stated that the Emergency Plan will describe (i) the emergency response organization and the responsibilities and authorities of key positions within the organization, (ii) the responsibilities of repository personnel during a radiological incident, (iii) positions within the organization that have the responsibility for declaring emergencies during normal hours when key personnel and shift organization are present and during off-normal hours when only shift organization is present, (iv) methods for activating the staff necessary for implementation of the Emergency Plan, and (v) the positions responsible for overall direction of emergency response and notification of local agencies and NRC during normal and off-normal hours.

In SAR Section 5.7.1.2, DOE stated that the site protection manager is a key manager who will be located onsite, will report to the site operations manager, and will be responsible for developing and maintaining the emergency preparedness program. DOE stated that the records will be maintained for reviews and updates of the Emergency Plan, for notification of repository personnel and other onsite or offsite response organizations affected by an update of the Emergency Plan or its implementing procedures, and for review and acceptance of the Emergency Plan or implementing procedure updates.

8.   Notification and Coordination of Offsite Groups

In SAR Section 5.7.8.1, DOE stated that the Emergency Plan will describe the Technical Support Center, located at the repository, and the Operations Facility Center, located offsite, as well as the organization responsible for activating the emergency response organization for performing timely notifications under accident conditions during normal and off-normal hours. Additionally, DOE stated that the Emergency Plan will identify the offsite location and describe the functions of the Joint Information Center. DOE stated that the Emergency Plan will also describe the means to notify offsite response organizations and the means to request offsite assistance, including medical assistance. Furthermore, responsible offsite agencies will be identified in the Emergency Plan and notification methods and equipment will be described and will be sufficiently diverse to ensure that notification and activation can be performed even if some personnel, equipment, or parts of the Emergency Operations Center are unavailable.

In SAR Section 5.7.8.2, DOE stated that emergency personnel will take actions as specified in the Emergency Plan and its implementing procedures. These actions include classifying the emergency, directing staff to assume emergency response roles, sounding the site emergency signal, and providing timely notification to appropriate Federal, State of Nevada, and local agencies of the emergency. The implementing procedures will contain the telephone, fax, address, and e-mail information necessary to achieve timely notification of responsible offsite agencies and points of contact. Additionally, DOE stated that the Emergency Plan will include a provision to contact the NRC Operations Center upon completion of local notifications but not later than 1 hour after an alert or site area emergency has been declared.

9.   Information To Be Communicated

In SAR Section 5.7.9.1, DOE stated that the Emergency Plan will describe the types of information to be provided on repository status and radioactive releases and any recommended protective actions that will be communicated to the offsite response organizations and to NRC in the event of an emergency. SAR Section 5.7.9.2 stated that this information will be consistent with the NRC Event Notification Worksheet (NRC Form 361) modified for repository use (as shown in SAR Figure 5.7-2). Furthermore, in SAR Section 5.7.10.2, DOE stated that the Emergency Plan will define the requirements to train repository personnel on how to respond to emergencies.

10.   Training

In SAR Section 5.7.10.1, DOE stated that the Emergency Plan will define the requirements to train repository personnel on how to respond to emergencies and describe any special instructions and orientation tours in the training offered to offsite support personnel, including police, fire, and medical personnel who may be called upon to respond in an emergency.

As stated in SAR Section 5.7.10.2, SAR Section 5.3 provided a general description of the organizational structure, as it is anticipated to exist at the time of repository construction and operations; the key positions assigned responsibility for safety and operations; and the personnel qualifications and training program. DOE (2008ai) stated that (i) the Emergency Preparedness Training Program document, developed as described in SAR Section 5.3.3, will contain the training requirements for each position in the emergency response organization, frequency of retraining, and estimated number of hours of initial training and retraining and (ii) personnel who are not members of the emergency response staff who are permitted inside the GROA without a full-time escort are required to receive General Employee Training, as discussed in SAR Section 5.3.3.

11.   Restoration of Repository Operations to a Safe Condition

In SAR Section 5.7.11.1, DOE stated that the Emergency Plan will contain a description of the means for restoring the repository to a safe condition after an emergency in accordance with recovery procedures, as well as criteria for the return to operations. Should the response result in the evacuation of areas, DOE stated that it will provide criteria for safe reentry. Additionally, DOE stated that the Emergency Plan will describe the procedures for restoring the facility to a safe status after an accident using recovery plans. DOE also stated that the recovery plans will include requirements for checking and restoring to normal operation all safety equipment important to safety, and requirements for returning emergency equipment and supplies used during an accident to a state of readiness.

12.   Exercises, Communication Checks, and Drills

In SAR Section 5.7.12.1, DOE stated that the Emergency Plan will describe the drills and exercises that will be used to evaluate the major portions of the emergency response capabilities and to maintain key response skills. In SAR Sections 5.7.12.2.1–5.7.12.2.10, DOE described an exercise program that includes quarterly communication checks with offsite response organizations; biennial exercises in the form of simulated emergencies; and semiannual radiological and health physics, medical, and fire drills.

13.   Hazardous Materials

In SAR Section 5.7.13.1, DOE stated that the Emergency Plan will include a certification that the repository has complied with the Emergency Planning and Community Right-to-Know Act of 1986 with respect to hazardous materials within the GROA. SAR Table 5.7-9 listed chemicals likely to be used at the repository.

14.   Comments on the Emergency Plan

In SAR Section 5.7.14.1, DOE stated that the Emergency Plan will be provided to offsite response organizations identified in the Emergency Plan for review prior to submittal to NRC. Additionally, the offsite response organizations will have 60 days to review and comment on the Emergency Plan, and any offsite response organization comments received will be included with the Emergency Plan that will be submitted to NRC. Furthermore, DOE stated that comments from offsite response organizations, as appropriate, will be dispositioned in subsequent revisions to the Emergency Plan. If subsequent revisions to the Emergency Plan affect the offsite response organizations, DOE stated that future revisions will also be provided to those organizations for review. DOE explained that the comment period for subsequent revisions to the Emergency Plan will be 60 days and any additional comments offsite organizations provide during this period will again be included with the revised Emergency Plan submitted to NRC.

15.   Offsite Assistance

In SAR Section 5.7.15.1, DOE stated that to facilitate a coordinated and planned emergency response, provisions for advance arrangements with offsite organizations will be addressed in the Emergency Plan. DOE's provisions for advance arrangements include (i) means for requesting offsite assistance and provisions that exist for using other organizations capable of augmenting the planned onsite response; (ii) provisions for prompt communications among principal response organizations to offsite emergency personnel who would be responding onsite; (iii) emergency facilities and equipment to support the emergency response; (iv) availability of methods, systems, and equipment for assessing and monitoring actual or potential consequences of a radiological emergency condition; (v) arrangements for medical services for contaminated and injured onsite individuals; and (vi) radiological emergency response training for those offsite support organizations that may be called upon to assist in an emergency onsite. DOE (2008ai) stated that the Emergency Plan will include a DOE offer to meet at least annually with each offsite response organization to review items of mutual interest, including relevant changes to the Emergency Plan, and will discuss the EAL scheme, notification procedures, and overall response coordination process during these meetings.

16.   Public Information

In SAR Section 5.7.16.2.1, DOE stated that the Repository Emergency Public Information Program establishes the means for providing accurate and timely information to workers on the repository site and the general public through the media. In addition, DOE stated that the Emergency Plan will describe arrangements for providing timely information to the public in the event of an emergency. This information will be disseminated to the media and the public through the Joint Information Center. DOE stated that in conjunction with state and local agencies, the repository will conduct an annual orientation for local news media to acquaint them with information that assists them in providing informed coverage of an event and lessen the possibility of errors in reporting.

NRC Staff Evaluation

On the basis of the NRC staff evaluation of SAR Section 5.7 and DOE's response to the NRC staff request for additional information (DOE, 2008ai), the NRC staff notes that DOE's description of its Emergency Plan is reasonable because DOE has provided a detailed description of the contents of the Emergency Plan, including

- A description of the GROA and the surrounding area

- Each type of radioactive materials accidents (NRC staff notes that DOE's use of the event sequence analysis, as part of the preclosure safety analysis, could enhance the description of the accidents in terms of the accident type including possible onsite and offsite consequences)

- Classification system to identify accidents as "alerts" or "site area emergencies"

- Means that DOE stated it will use to detect key initiating events and accident conditions

- Description of the means to mitigate consequences of each accident type, including the means to protect site workers and to maintain mitigation equipment

- Description of methods and equipment to assess releases of radioactive materials

- Description of personnel responsibilities should an accident occur, including the responsibilities for developing, maintaining, and updating the emergency plan

- Description of the means to notify the offsite response organizations

- Description of the types of information to be provided on facility status, radioactive releases, and recommended protective actions that are to be provided to offsite response organizations

- Description of the training for site workers on how to respond to an emergency and any special instructions and orientation tours offered to fire, police, medical, and other offsite-based emergency personnel

- Description of the means to restore the facility to a safe condition after an accident

- Quarterly communication checks with offsite response organizations; biennial onsite exercises to test responses to simulated emergencies; and semiannual radiological/health physics, medical and fire drills

- Certification for the Community Right-to-Know Act of 1986 regarding hazardous materials

- Description of the approach for obtaining comments on the Emergency Plan from offsite response organizations

- Plans for the use of offsite assistance

- Arrangements for providing timely information to the public

2.5.7.4   NRC Staff Conclusions

Although a detailed plan is not available at this time, the NRC staff notes that DOE's description of its plan for responding to, and recovering from, radiological emergencies that may occur any time before permanent closure and decontamination or decontamination and dismantlement of surface facilities is reasonable, in light of the information available.

2.5.7.5   References

DOE. 2009av. DOE/RW–0573, "Yucca Mountain Repository License Application." Rev. 1. ML090700817. Las Vegas, Nevada: DOE, Office of Civilian Radioactive Waste Management.

DOE. 2008ab. DOE/RW–0573, "Yucca Mountain Repository License Application." Rev. 0. ML081560400. Las Vegas, Nevada: DOE, Office of Civilian Radioactive Waste Management.

DOE. 2008ai. "Yucca Mountain—Response to Request for Additional Information Regarding License Application (Safety Analysis Report Section 5.7), Safety Evaluation Report Vol. 4, Chapter 2.5.7, Set 1." Letter (November 5) B.J. Benney to J.R. Williams (NRC). ML083110290. Washington, DC: DOE, Office of Technical Management.

NRC. 2003aa. NUREG–1804, "Yucca Mountain Review Plan—Final Report." Rev. 2.
ML032030389. Washington, DC: NRC.

NRC. 2000aa. ISG–16, "Emergency Planning." Rev. 0. Washington, DC: NRC, Spent Fuel
Project Office.

NRC. 2000ab. NUREG–1567, "Standard Review Plan for Spent Fuel Dry Storage Facilities."
Washington, DC: NRC, Spent Fuel Project Office.

CHAPTER 11

2.5.8 Controls To Restrict Access and Regulate Land Uses

2.5.8.1 Introduction

This chapter evaluates the description provided in the U.S. Department of Energy (DOE) Safety Analysis Report (SAR) Section 5.8 (DOE, 2008ab, 2009av) of controls to restrict access and regulate land uses at the Yucca Mountain site and adjacent areas, and DOE's response to a U.S. Nuclear Regulatory Commission (NRC) request for additional information (RAI) (DOE, 2009au).

2.5.8.2 Evaluation Criteria

10 CFR 63.21(c)(24) requires DOE to include a description of the controls to restrict access and to regulate land uses at the Yucca Mountain site and adjacent areas, including a conceptual design of monuments that would be used to identify the site after permanent closure. 10 CFR 63.121, Requirements for Ownership and Control of Interests in Land, is divided into four parts: ownership of the land where the geologic repository operations area (GROA) is located [10 CFR 63.121(a)(1) and (2)], additional controls through permanent closure [10 CFR 63.121(c)], additional controls for permanent closure [10 CFR 63.121(b)], and water rights [10 CFR 63.121(d)(1) and (2)]. NRC staff used applicable guidance in the Yucca Mountain Review Plan (YMRP) Section 2.5.8 (NRC, 2003aa). The acceptance criteria follow:

- Ownership of land is adequately demonstrated
- Additional controls for permanent closure are acceptable
- Additional controls through permanent closure are adequate
- Description of water rights is adequate
- Conceptual design of monuments is adequate

2.5.8.3 Technical Evaluation

The GROA is a high-level radioactive waste (HLW) facility that is part of a geologic repository, including both surface and subsurface areas, where waste handling activities are conducted. The evaluation that follows is presented in five subsections: (i) ownership of land, (ii) additional controls for permanent closure, (iii) additional controls through permanent closure, (iv) water rights, and (v) conceptual design of monuments.

1. Ownership of Land

DOE provided a description of the proposed GROA, using Public Land Survey System nomenclature (i.e., township, range, and section). In SAR Section 5.8.1, DOE stated that the GROA and surrounding land within its proposed land withdrawal boundary are currently under control of several different Federal agencies, including DOE, the U.S. Department of the Interior, and the U.S. Department of Defense. DOE described in SAR Section 5.8.1 the course of action it pursued with respect to ownership of land. A land withdrawal bill was submitted to Congress in 2007 for the GROA and surrounding area (Senate Bill S.37, introduced May 23, 2007, in the 110th Congress). The proposed land withdrawal bill was not enacted into law, and DOE did not assert or demonstrate in the SAR that it has acquired by other means the land where the GROA would be located. DOE does not possess any legal interests that authorize the construction and operation of a repository (i.e., existing interests are limited in duration and are designated for other purposes, such as site characterization), but DOE stated that the GROA will be located in and on lands that are either acquired lands under the jurisdiction and control of DOE or will be permanently withdrawn and reserved for its use.

2. Additional Controls for Permanent Closure

DOE described its approach for controls for permanent closure in SAR Section 5.8.2. DOE identified controls that will be implemented for permanent closure including site monuments and markers, public records and archives, and government ownership and regulations for land or resource use. DOE identified on a map in SAR Figure 5.8-2 (DOE, 2008ab) the proposed postclosure controlled area boundary and proposed land ownership boundary. In response to an NRC staff request for additional information (RAI), DOE (2009au) clarified the proposed postclosure controlled area boundary and provided a description of this area.

In SAR Section 5.8.2, DOE stated that it will exercise jurisdiction and control over surface and subsurface domains necessary to prevent adverse human actions that could significantly reduce the ability of the repository to achieve waste isolation and that appropriate controls necessary to prevent such adverse human actions will be implemented at the repository. Access controls would have DOE control or own the land on which it would apply the access controls. SAR Figure 5.8-2 depicts that the postclosure controlled area is wholly contained within the proposed land withdrawal boundary.

In SAR Section 5.8.2.3, DOE described its approach for administering and controlling ownership rights for permanent closure. DOE (2009au) clarified its administrative program that identifies and defines any restrictions and controls for land areas outside of the GROA. DOE stated it would submit draft license specifications to NRC prior to the NRC issuance of the license to receive and possess spent nuclear fuel and HLW. The access control program will also address any restrictions or controls needed during the preclosure period to ensure postclosure safety. As described in SAR Table 5.10-3, the access control program would be modified prior to permanent closure of the repository to address any continuing or additional restrictions or controls for the postclosure period. DOE stated that it would identify continuing or additional restrictions or controls needed after permanent closure, and these would be identified prior to permanent closure.

3. Additional Controls Through Permanent Closure

Additional controls through permanent closure (i.e., controls for the preclosure period) are related to DOE jurisdiction and control of activities necessary to ensure safety. DOE identified on a map in SAR Figure 5.8-2 (DOE, 2008ab) the proposed preclosure controlled area boundary, which extends beyond the GROA. This map also shows that the boundaries of the proposed preclosure controlled area and the proposed land ownership boundary from the land withdrawal bill are the same. DOE (2009au) clarified the proposed preclosure controlled area boundary relative to the ownership boundary identified for land withdrawal and provided a legal description of this area, which defined the preclosure controlled area boundary and showed that an identified plot of land (Patent 27-83-002) within this boundary would remain under private ownership. DOE (2009au) noted the size of Patent 27-83-002 {73.9 ha [182.5 acre]} and that this privately owned plot (mining claim property) would not be part of the land withdrawn and is not part of the controlled area over which DOE controls access. This excluded plot is located near the southwestern edge of the preclosure controlled area, about 16.1 km [10 mi] from the GROA. DOE (2009au) stated it would update figures in the SAR to indicate the correct acreage of Patent 27-83-002 and that the mining claim property is excluded from the proposed land withdrawal boundary and is not part of the controlled area.

In SAR Section 5.8.3.1, DOE described that the consequence analysis performed in SAR Section 1.8 considered the size and boundaries of the GROA and the site, and the locations of the highest airborne concentrations in the general environment and in areas outside the general environment. On the basis of its consequence analysis, in SAR Section 5.8.3, DOE identified two types of controls it would apply to the preclosure controlled area (flight restrictions to help ensure preclosure safety and access control to prevent disturbance of the GROA).

DOE stated that the GROA will be located in and on lands that are either acquired lands under the jurisdiction and control of DOE or will be permanently withdrawn and reserved for its use. In addition, DOE has stated that the proposed preclosure controlled area is coincident with the land withdrawal boundary.

DOE described a phased approach for addressing and managing restrictions and controls for activities outside the GROA, but within the site boundary. In response to an NRC staff RAI, DOE (2009au) clarified its administrative program that identifies and defines any restrictions and controls for land areas outside of the GROA. To reflect this approach, DOE (2009au) stated it would revise the SAR Section 5.8.3 to state that an access control program will be implemented to ensure preclosure safety.

4.   Water Rights

In SAR Section 5.8.4 and DOE (2009au), DOE described its approach for water rights. DOE estimated a maximum annual water demand of 53.0 hectare-meters [430 acre-ft] for construction (prior to receipt and possession of spent nuclear fuel and HLW) and a maximum annual water demand of 40.7 hectare-meters [330 acre-ft] for operations (after receipt and possession of spent nuclear fuel and HLW). DOE has filed a water appropriation request with the State of Nevada for the permanent rights to 53.0 hectare-meters [430 acre-ft] annually from five wells within the proposed preclosure controlled area boundary for the purpose of constructing and operating the repository. On July 22, 1997, the Nevada State Engineer denied the DOE water appropriation permit applications. DOE stated that the U.S. Department of Justice, on behalf of DOE, appealed this decision in U.S. District Court.

Water rights are included in the additional controls to be established for permanent closure because potential postclosure uses of water by others could significantly reduce the geologic repository's ability to achieve waste isolation. DOE (2009au) stated that it believes no additional water rights controls are needed for permanent closure; however, it intends to develop an access control plan that will include any controls that may be necessary for permanent closure. These controls, which DOE stated will be modified from existing preclosure controls or developed as part of the determination to close the repository, will assure that it will evaluate access to the postclosure controlled area following closure (and any related activities, such as drilling) to determine the impact to the postclosure safety analyses prior to allowing such access. DOE stated that it will evaluate any existing wells and associated water rights at the time of permanent closure as part of these access controls and will take any actions needed at that time.

5.   Conceptual Design of Monuments

SAR Section 5.8.5 addresses DOE's conceptual design of monuments. DOE described that
(i) monuments and markers identify the GROA, postclosure controlled area, and preclosure
controlled area; (ii) the conceptual design includes monument design, fabrication, and
emplacement considerations to ensure monuments are as permanent as practicable; and
(iii) the monuments will communicate information and warnings. DOE stated that it will submit,
prior to permanent closure, a detailed description of the measures to be employed, such as
construction of monuments, to regulate or prevent activities that could impair the long-term
isolation of emplaced waste within the geologic repository and to assure that relevant
information will be preserved for the use of future generations.

NRC Staff Evaluation

On the basis of its review of the DOE information (SAR Section 5.8), NRC staff notes that,
although the GROA is not located on lands that are under the jurisdiction and control of DOE or
lands that have been permanently withdrawn and set aside for its use, DOE took steps within its
purview by proposing a land withdrawal bill in 2007. DOE has not provided legal documentation
for acquisition or withdrawal of this land, but stated that in the case of a legislative withdrawal, a
citation to the legislation, with inclusion of pertinent provisions of the legislation, will be included
in a revision to the license application, and in the case of other land acquisition activities, a
sufficient index of ownership and control will be available to satisfy a purchaser of record.
Additionally, DOE has taken steps within its purview to obtain water rights, but has not yet
obtained ownership of the water rights. DOE explained that the U.S. Department of Justice, on
behalf of DOE, continues to appeal the State of Nevada Engineer's decision to deny DOE's
permit for its water demands.

Regarding DOE's other descriptions of the controls to restrict access and regulate land uses,
the NRC staff notes that DOE's description of additional controls for permanent closure,
additional controls through permanent closure, and the conceptual design of monuments are
reasonable because of the following:

- DOE has defined the GROA at a reasonable level of detail using Public Land Survey
  System nomenclature.

- DOE identified the proposed boundary of the preclosure controlled area and identified
  two types of controls (flight restrictions and access control).

- DOE described its approach for controls for permanent closure, which include site
  monuments and markers, public records and archives, and government ownership and
  regulations for land and resources.

- DOE described the conceptual design of the monuments and markers that would be
  used to identify the site after permanent closure.

2.5.8.4   NRC Staff Conclusions

On the basis of its review, the NRC staff notes that DOE's description of the controls to restrict
access and regulate land uses is reasonable. The DOE's description of controls with respect to
(i) additional controls for permanent closure, (ii) additional controls through permanent closure,

and (iii) the conceptual design of monuments is consistent with the guidance in the YMRP.   The NRC staff notes that DOE has taken the steps within its purview to obtain ownership of land and water rights; however, at this time, DOE does not own the land or water rights.

2.5.8.5　　　References

DOE. 2008ab. DOE/RW–0573, "Safety Analysis Report Yucca Mountain Repository License Application." Rev. 0. ML081560400. Las Vegas, Nevada: DOE, Office of Civilian Radioactive Waste Management.

DOE. 2009au. "Yucca Mountain—Response to Request for Additional Information Regarding License Application (Safety Analysis Report Section 5.8), Safety Evaluation Report Vol. 4, Chapter 2.5.8, Set 1." Letter (May 6) J.R. Williams to F. Jacobs (NRC). ML091330698. Washington, DC: DOE, Office of Technical Management.

DOE. 2009av. DOE/RW–0573, "Safety Analysis Report Yucca Mountain Repository License Application." Rev. 1. ML090700817. Las Vegas, Nevada: DOE, Office of Civilian Radioactive Waste Management.

NRC. 2003aa. NUREG–1804, "Yucca Mountain Review Plan—Final Report." Rev. 2. ML032030389. Washington, DC: NRC.

CHAPTER 12

2.5.9 Uses of Geologic Repository Operations Area for Purposes Other Than Disposal of Radioactive Wastes

2.5.9.1 Introduction

This chapter evaluates the description of the plans provided in the U.S. Department of Energy (DOE) Safety Analysis Report (SAR) Section 5.9 (DOE, 2008ab, 2009av) for uses of the geologic repository operations area (GROA) for purposes other than disposal of radioactive wastes. DOE provided further information in its response to an NRC staff request for additional information (DOE, 2008ad).

2.5.9.2 Evaluation Criteria

10 CFR 63.21(c)(22)(vii) requires the SAR to include plans for any uses of the GROA for purposes other than radioactive waste disposal, with an analysis of the effects, if any, that such uses may have on the operation of the structures, systems, and components (SSCs) important to safety and the engineered and natural barriers important to waste isolation. The NRC staff used applicable guidance in the Yucca Mountain Review Plan (YMRP) Section 2.5.9 (NRC, 2003aa) in its review of this information. The acceptance criteria from the YMRP with respect to uses of the GROA for purposes other than disposal of radioactive wastes are (i) the proposed activities other than disposal of radioactive wastes are acceptable and (ii) the procedures for proposed activities other than disposal of radioactive wastes are acceptable.

2.5.9.3 Technical Evaluation

This evaluation is divided into (i) proposed uses for purposes other than disposal, and effects on performance and (ii) procedures for proposed activities other than high-level waste disposal.

1. Proposed Uses for Purposes Other Than Waste Disposal, and Effects on Performance

SAR Section 5.9.1 identified three types of activities that it considered for potential other uses of the GROA: Native American cultural activities, independent performance monitoring, and flora and fauna protection. Native American cultural activities include (i) ongoing activities to protect cultural artifacts and other resources and (ii) any present or future utilization of the GROA or repository site for ceremonial or other cultural purposes. The SAR also described other possible activities that are expressly expected not to occur in the GROA; specifically, exploitation of geothermal, mineral, metal, or water resources. In addition, the introduction to SAR Section 5.9 stated that no long-term interim waste storage is planned as part of repository operations, although aging incident to disposal is expected. Aging incident to disposal is part of waste disposal operations, is not within the scope of this section, and is considered in the TER Preclosure Volume.

DOE stated that it will authorize other uses of the GROA only after performing an analysis demonstrating that the associated activities will have no adverse effect on SSCs important to safety or performance of barriers important to waste isolation. DOE described the activities and possible impacts for continuations of two present activities: cultural resource protection, and flora and fauna protection. DOE acknowledged that it may receive requests for use of the GROA from Native Americans for ceremonial or other cultural heritage purposes and from groups other than NRC and DOE to conduct performance monitoring. DOE did not describe specific activities or provide any analyses of potential effects on repository operations for ceremonial or other cultural heritage purposes of Native Americans, or performance confirmation from groups other than NRC or DOE. DOE stated it would develop procedures to evaluate such requests prior to the handling of waste and that the procedures will include evaluation of the purpose of the activity, detailed activity descriptions for evaluation, radiation safety of workers and visitors, and disposition of records and identification of parties to be notified upon completion of the activities.

DOE (2008ad) stated that surface-disturbing activities associated with cultural resource protection could potentially affect the operation of SSCs important to safety and the engineered and natural barriers important to waste isolation. DOE determined that surface-disturbing activities associated with cultural resource protection are similar but less challenging to the operation of SSCs important to safety than the repository construction activities analyzed in SAR Section 1.6.3.5, where DOE determined that construction activities do not initiate event sequences that could pose a threat to the operation of SSCs important to safety. DOE determined that the engineered barrier system is not affected by surface-disturbing activities, as this system's features are located underground, far below any ground surface activity.

DOE analyzed potential impacts from cultural resource activities to three areas of the natural system: the topography and surficial soils of the upper natural barrier, the unsaturated zone, and the saturated zone. DOE (2008ad) stated that effects from surface-disturbing activities on the function of the topography and surficial-soils features of the upper natural barrier are controlled by Design Control Parameter 09-04 (SAR Table 1.9.8), such that the function of these features is maintained. DOE stated that surface-disturbing activities could affect the unsaturated zone if infiltration increased and the saturated zone if recharge increased. DOE's analysis indicated that these impacts would be limited in both magnitude and time, given the limited extent of the surface-disturbing activities relative to the overall size of the GROA and the limited time of disturbance prior to restoration and reclamation.

DOE (2008ad) determined that reclamation activities associated with flora and fauna protection could potentially affect operation of SSCs important to safety and the engineered and natural barriers important to waste isolation. DOE's impact analysis for flora and fauna protection activities on the operation of SSCs important to safety and the engineered and natural barriers important to waste isolation is consistent with its cultural resource protection analysis. DOE reached the same conclusions and provided the same basis for reaching this determination.

The NRC staff notes that the DOE impact analyses are reasonable for activities currently identified for cultural resource protection and flora and fauna protection because the DOE analyses (i) considered the principal impact of land disturbance and (ii) determined that for the operation of SSCs important to safety, these reclamation and other surface disturbances will be minor in scale and extent compared to construction-related disturbances, which include such activities as excavation, tunnel boring, drilling, and blasting. Additionally, the NRC staff notes that these reclamation and other surface disturbances would have a minimal impact on the natural and engineered barriers important to waste isolation because of the (i) location of the activities relative to the waste disposal location, (ii) procedural controls implemented to maintain operation of the barriers, and (iii) limited magnitude and duration of these reclamation and surface disturbance activities.

2. Procedures for Proposed Activities Other Than Waste Disposal

DOE detailed procedures to manage the two ongoing other uses of the GROA: protection of cultural resources and protection of flora and fauna. DOE stated that these procedures were implemented in part to fulfill DOE's responsibilities under the National Historic Preservation Act of 1966, Sections 106 and 110 (16 U.S.C. 470 et seq.), and the Endangered Species Act of 1973 (16 U.S.C. 1531 et seq.).

DOE identified two other expected future uses of the GROA: ceremonial or other cultural heritage purposes, and independent performance monitoring. DOE stated that it will develop procedures, prior to waste handling, for authorizing requests to use the GROA for purposes other than waste disposal. DOE stated that these procedures will require a disclosure of the purpose of the activity, a detailed activity description for evaluation, an analysis demonstrating that the activities will have no adverse effect on SSCs or features that are important to safety or important to waste isolation, and access authorization procedures for individuals and groups engaged in these activities.

The NRC staff notes that DOE's description of its procedures for proposed activities other than waste disposal is reasonable because the planned procedures (i) address the relevant activities, including expected future activities; (ii) identify relevant statutes (i.e., the Historic Preservation Act and Endangered Species Act) that would be used to support identification of activities and analysis of impacts from these uses; and (iii) the procedures include analyses to evaluate the activities' effect on SSCs or features that are important to safety or important to waste isolation.

2.5.9.4 NRC Staff Conclusions

On the basis of its review, the NRC staff notes that the descriptions of the plans and procedures for activities other than waste disposal are reasonable. The DOE's description of its plans for other uses is consistent with the guidance in the YMRP.

2.5.9.5 References

DOE. 2009av. DOE/RW–0573, "Yucca Mountain Repository License Application." Rev. 1. ML090700817. Las Vegas, Nevada: DOE, Office of Civilian Radioactive Waste Management.

DOE. 2008ab. DOE/RW–0573, "Yucca Mountain Repository License Application." Rev. 0. ML081560400. Las Vegas, Nevada: DOE, Office of Civilian Radioactive Waste Management.

DOE. 2008ad. "Yucca Mountain—Response to Request for Additional Information Regarding License Application (Safety Analysis Report Section 5.9), Safety Evaluation Report Vol. 4, Chapter 2.5.9, Set 1." Letter (October 22) J.R. Williams to B. Benney (NRC). ML082960784. Las Vegas, Nevada: DOE, Office of Civilian Radioactive Waste Management.

NRC. 2003aa. NUREG–1804, "Yucca Mountain Review Plan—Final Report." Rev. 2. ML032030389. Washington, DC: NRC.

CHAPTER 13

Conclusions

The U.S. Nuclear Regulatory Commission (NRC) staff reviewed the Safety Analysis Report (SAR) and the other information submitted by the U.S. Department of Energy (DOE) and notes that the information provided by DOE is reasonable. More specifically, the NRC staff notes the following:

- On the basis of the evaluation in Chapter 1, a research and development program is not needed at this time (neither DOE in its SAR nor the NRC staff in its review of the SAR has identified any safety questions for the research and development program). DOE's description of its approach for developing and implementing a research and development program is reasonable as a general approach for use in addressing safety questions should a safety question be identified in the future.

- On the basis of the evaluation in Chapter 2, DOE has provided a reasonable description of its Performance Confirmation Program that is consistent with the guidance in the Yucca Mountain Review Plan (YMRP).

- On the basis of the evaluation in Chapter 3, DOE's description of its quality assurance program, including the Office of Civilian Radioactive Waste Management (OCRWM) Quality Assurance Requirements and Description (QARD), is reasonable and consistent with the guidance in the YMRP.

- On the basis of the evaluation in Chapter 4, DOE's descriptions of programs for records, reports, tests, and inspections are reasonable and consistent with the guidance in the YMRP.

- On the basis of the evaluation in Chapter 5, DOE has provided a reasonable description of the organizational structure for the construction and operation of the geologic repository operations area (GROA), including a description of any delegations of authority and assignments of responsibilities. The DOE's organizational structure is consistent with the guidance in the YMRP.

- On the basis of the evaluation in Chapter 6, DOE reasonably described the key positions assigned responsibility for GROA safety and operations and the qualifications of the persons occupying these positions. The DOE's description of the key positions and qualifications (i.e., minimum skills and experience) is consistent with the guidance in the YMRP, which recognizes that DOE is not expected to have identified specific individuals to fill key positions for a construction authorization.

- On the basis of the evaluation in Chapter 7, DOE provided reasonable information on a personnel qualifications and training program that DOE stated will be implemented before DOE receives, possesses, stores, or disposes high-level radioactive waste. The DOE's information regarding its personnel qualifications and training program is consistent with the guidance in the YMRP, which recognizes that DOE is not expected to have an NRC-approved personnel qualifications and training program in place for a construction authorization.

- On the basis of the evaluation in Chapter 8, DOE has reasonably described the plans for startup activities and testing that DOE stated will be implemented before DOE receives, processes, stores, or disposes high-level radioactive waste. The DOE's description of its plans for startup activities and testing is consistent with the guidance in the YMRP, which recognizes that DOE is not expected to have prepared plans for startup activities and testing for a construction authorization.

- On the basis of the evaluation in Chapter 9, DOE reasonably described plans for conduct of normal activities, including maintenance, surveillance, and periodic testing that DOE stated will be implemented before receipt, possession, storing, or disposal of high-level radioactive waste. The DOE's description of its plans for normal activities, including maintenance, surveillance, and periodic testing, is consistent with the guidance in the YMRP, which recognizes that DOE is not expected to have plans in place for normal activities for a construction authorization.

- Although a detailed emergency plan is not available at this time, on the basis of the evaluation in Chapter 10, DOE's description of its plan for responding to, and recovering from, radiological emergencies that may occur any time before permanent closure and decontamination or decontamination and dismantlement of surface facilities is reasonable, in light of the information available.

- On the basis of the evaluation in Chapter 11, DOE's description of the controls to restrict access and regulate land uses is reasonable. DOE's description of controls with respect to (i) additional controls for permanent closure, (ii) additional controls through permanent closure, and (iii) the conceptual design of monuments is consistent with the guidance in the YMRP. The NRC staff notes that DOE has taken the steps within its purview to obtain ownership of land and water rights; however, at this time, DOE does not own the land or water rights.

- On the basis of the evaluation in Chapter 12, DOE's descriptions of the plans and procedures for activities other than waste disposal are reasonable. The DOE's description of its plans for other uses is consistent with the guidance in the YMRP.

CHAPTER 14

Glossary

This glossary is provided for information and is not exhaustive. Terms shown in *italics* are included in this glossary.

absorption: The process of taking up by capillary, osmotic, solvent, or chemical action of molecules (e.g., absorption of gas by water) as distinguished from *adsorption*.

adsorption: The adhesion by chemical or physical forces of molecules or ions (as of gases or liquids) to the surface of solid bodies. For example, the transfer of solute mass, such as *radionuclides*, in *groundwater* to the solid geologic surfaces with which it comes in contact. The term *sorption* is sometimes used interchangeably with this term.

advection: The process in which solutes, particles, or molecules are transported by the motion of flowing fluid.

aging: The retention of *commercial spent nuclear fuel* on the surface in *dry storage* to reduce its thermal output as necessary to meet repository thermal management goals.

Alloy 22: A nickel-based, *corrosion*-resistant alloy containing approximately 22 weight percent chromium, 13 weight percent molybdenum, and 3 weight percent tungsten as major alloying elements. This alloy is used as the outer container material in DOE's waste package design.

alluvium: Detrital (sedimentary) deposits made by flowing surface water on river beds, flood plains, and alluvial fans. It does not include subaqueous sediments of seas and lakes.

ambient: Undisturbed, natural conditions, such as ambient temperature caused by climate or natural subsurface thermal gradients, and other surrounding conditions.

anisotropy: Variation in physical properties when measured in different directions. For example, in layered rock, permeability is often greater within the horizontal layers than across the horizontal layers.

aqueous: Pertaining to water, such as aqueous phase, aqueous species, or aqueous *transport*.

aquifer: A saturated underground geologic formation of sufficient permeability to transmit *groundwater* and yield water of sufficient quality and quantity to a well or spring for an intended beneficial use.

calibration: (1) Comparison of model results with actual data or observations, and adjusting model parameters to increase the precision and/or accuracy of model results compared to actual data or observations. (2) For tools used for field or lab measurements, the process of taking instrument readings on standards known to produce a certain response, to check the accuracy and precision of the instrument. (3) In operations, the process to ensure accuracy of instruments and any setpoints for automation actuations of items important to safety.

Category 1 event sequences: Those event sequences that are expected to occur one or more times before permanent closure of a proposed geologic repository.

Category 2 event sequences: Event sequences other than Category 1 event sequences that have at least one chance in 10,000 of occurring before permanent closure.

colloid: As applied to *radionuclide* migration, colloids are large molecules or very small particles, having at least one dimension with the size range of 10^{-6} to 10^{-3} mm [10^{-8} to 10^{-5} in] that are suspended in a solvent. Colloids in *groundwater* arise from clay minerals, organic materials, or (in the context of a geologic repository) from corrosion of engineered materials.

commercial spent nuclear fuel: Nuclear fuel rods, forming a fuel assembly, that have been removed from a nuclear power plant after reaching the specified *burnup*.

conceptual model: A set of qualitative assumptions used to describe a system or subsystem for a given purpose. Assumptions for the model are compatible with one another and fit the existing data within the context of the given purpose of the model.

consequence: A measurable or calculated outcome of an event or process that, when combined with the probability of occurrence, gives a measurement of *risk*.

corrosion: The deterioration of a material, usually a metal, as a result of a chemical or electrochemical reaction with its environment.

coupled processes: A representation of the interrelationships between *processes* such that the effects of variation in one process are accurately propagated among all interrelated *processes*.

crevice corrosion: *Localized corrosion* of a metal surface at, or immediately adjacent to, an area that is shielded from full exposure to the environment because of close proximity between the metal and the surface of another material.

criticality: The condition in which a fissile material sustains a chain reaction. It occurs when the number of neutrons present in one generation cycle equals the number generated in the previous cycle. The state is considered critical when a self-sustaining nuclear chain reaction is ongoing.

diffusion: (1) The spreading or dissemination of a substance caused by concentration gradients. (2) The gradual mixing of the molecules of two or more substances because of random thermal motion.

dispersion (hydrodynamic dispersion): (1) The tendency of a solute (substance dissolved in *groundwater*) to spread out from the path it is expected to follow if only the bulk motion of the flowing fluid were to move it. The tortuous path the solute follows through openings (pores and fractures) causes part of the dispersion effect in the rock. (2) The macroscopic outcome of the actual movement of individual solute particles through a porous medium. Dispersion dilutes solutes, including *radionuclides*, in *groundwater*.

dissolution: Dissolving a substance in a solvent.

drift: From mining terminology, a horizontal underground passage. In the Yucca Mountain repository design, drifts include excavations for emplacement (emplacement drifts) and access (access mains).

drift degradation: The progressive accumulation of rock rubble in a *drift* created by weakening and collapse of drift walls in response to stress from heating or earthquakes.

drip shield: A metallic structure placed along the extension of the emplacement *drifts* and above the waste packages to prevent *seepage* water from directly dripping onto the waste package outer surface. The drip shield may also prevent the *drift* ceiling rocks (e.g., due to *drift degradation*) from falling on the waste package.

dry storage: Storage of *spent nuclear fuel* without immersion of the fuel in water for cooling or shielding; it involves the encapsulation of spent fuel in a steel cylinder that might be in a concrete or massive steel *cask* or structure.

emplacement drift: See *drift*.

events: In a *total system performance assessment*, (1) occurrences of phenomena that have a specific starting time and, usually, a duration shorter than the time being simulated in a *model*. (2) Uncertain occurrences of phenomena that take place within a short time relative to the time frame of the model.

exploratory studies facility: An underground laboratory at Yucca Mountain that includes a 7.9-km [4.9-mi] main loop (tunnel); a 2.8-km [1.75-mi] cross drift; and a research alcove system constructed for performing underground studies during site characterization.

fault (geologic): A planar or gently curved *fracture* across which there has been displacement parallel to the *fracture* surface.

features: Physical, chemical, thermal, or temporal characteristics of the site or potential repository system. For the purposes of screening features, *events*, and *processes* for the total system performance assessment, a feature is defined to be an object, structure, or condition that has a potential to affect disposal system performance.

flow: The movement of a fluid such as air, water, or *magma*. Flow and *transport* are processes that can move *radionuclides* from the proposed repository to the receptor group location.

fracture: A planar discontinuity in rock along which loss of cohesion has occurred. It is often caused by the stresses that cause folding and faulting. A *fracture* along which there has been displacement of the sides relative to one another is called a *fault*. A fracture along which no appreciable movement has occurred is called a joint. Fractures may act as fast paths for *groundwater* movement.

frequency: The number of occurrences of an observed or predicted event during a specific time period.

groundwater: Water contained in pores or *fractures* in either the unsaturated or saturated zones below ground level.

hydrologic: Pertaining to the properties, distribution, and circulation of water on the surface of the land, in the soil and underlying rocks, and in the atmosphere.

igneous: (1) A type of rock that has formed from a molten, or partially molten, material. (2) A type of activity related to the formation and movement of molten rock, either in the subsurface or on the surface.

infiltration: The process of water entering the soil at the ground surface. Infiltration becomes percolation when water has moved below the depth at which evaporation or *transpiration* can return it to the atmosphere. See also *net infiltration*.

invert: A constructed surface that would provide a level *drift* floor and enable emplacement and support of the waste packages.

lithophysal: Containing lithophysae, which are holes in *tuff* and other volcanic rocks. One way lithophysae are created is by the accumulation of volcanic gases during the formation of the tuff.

localized corrosion: Corrosion at discrete sites (e.g., pitting and *crevice corrosion*).

magma: Molten or partially molten rock that is naturally occurring and is generated within the earth. Magma may contain crystals along with dissolved gasses.

matrix: Rock material and its pore space exclusive of *fractures*. As applied to Yucca Mountain tuff, the ground mass of an *igneous* rock that contains larger crystals.

model: A depiction of a system, phenomenon, or process, including any hypotheses required to describe the system or explain the phenomenon or process.

near- field: The area and conditions within the potential repository including the *drifts* and waste packages and the rock immediately surrounding the *drifts*. The near-field is the region in and around the potential repository where the excavation of the repository *drifts* and the emplacement of waste have significantly impacted the natural *hydrologic* system.

net infiltration: The downward flux of infiltrating water that escapes below the zone of evapotranspiration. The bottom of the zone of evapotranspiration generally coincides with the lowermost extent of plant roots.

permeability: A measure of the ease with which a fluid such as water or air moves through a rock, soil, or sediment.

porosity: The ratio of the volume occupied by openings, or voids, in a soil or rock, to the total volume of the soil or rock. Porosity is expressed as a decimal fraction or as a percentage.

processes: Phenomena and activities that have gradual, continuous interactions with the system being modeled.

process model: A depiction or representation of a *process*, along with any hypotheses required to describe or to explain the *process*.

radiolysis: Chemical decomposition by the action of radiation.

radionuclide: An unstable isotope of an element that decays or disintegrates spontaneously, thereby emitting ionizing radiation. Approximately 5,000 natural and artificial radioisotopes have been identified.

risk: The probability that an undesirable event will occur, multiplied by the consequences of the undesirable event.

risk assessment: An evaluation of potential *consequences* or hazards that might be the outcome of an action, including the likelihood that the action might occur. This assessment focuses on potential negative impacts on human health or the environment.

risk informed, performance based: A regulatory approach in which *risk* insights, engineering analysis and judgments, and performance history are used to (i) focus attention on the most important activities; (ii) establish objective criteria on the basis of *risk* insights for evaluating performance; (iii) develop measurable or calculable parameters for monitoring system and licensee performance; and (iv) focus on the results as the primary basis for regulatory decision making.

rockfall: The release of *fracture*-bounded blocks of rock from the *drift* wall, usually in response to an earthquake.

scenario: A well-defined, connected sequence of *features, events*, and *processes* that can be thought of as an outline of a possible future condition of the potential repository system. Scenarios can be undisturbed, in which case the performance would be the expected, or nominal, behavior for the system. Scenarios can also be disturbed, if altered by disruptive events such as human intrusion or natural phenomena such as *volcanism* or nuclear *criticality*.

seepage: The inflow of *groundwater* moving in *fractures* or *matrix* pores of permeable rock to an open space in the rock. For the Yucca Mountain repository, seepage refers to water dripping into a *drift*.

seismic: Pertaining to, characteristic of, or produced by earthquakes or Earth vibrations.

sorption: The binding, on a microscopic scale, of one substance to another. Sorption is a term that includes both *adsorption* and *absorption* and refers to the binding of dissolved *radionuclides* onto geologic solids or waste package materials by means of close-range chemical or physical forces.   Sorption is a function of the chemistry of the radioisotopes, the fluid in which they are carried, and the material they encounter along the *flow* path.

source term: Types and amounts of *radionuclides* that are the source of a potential release.

spent nuclear fuel: Nuclear reactor fuel that has been used to the extent that it can no longer effectively sustain a chain reaction and that has been withdrawn from a nuclear reactor following irradiation, the constituent elements of which have not been separated by reprocessing. This fuel is more radioactive than it was before irradiation and releases significant amounts of heat from the decay of its fission product *radionuclides*.

stainless steel: A class of iron-base alloys containing a minimum of approximately 10 percent chromium to provide *corrosion* resistance in a wide variety of environments.

stress corrosion cracking: A cracking process that requires the simultaneous action of a corrosive substance and sustained (residual or applied) tensile stress. Stress corrosion cracking excludes both the fracture of already corroded sections and the localized corrosion processes that can disintegrate an alloy without the action of residual or applied stress.

structures, systems, and components: A *structure* is an element, or a collection of elements, that provides support or enclosure, such as a building, *aging* pad, or *drip shield*. A *system* is a collection of components, such as piping; cable trays; conduits; or heating, ventilation, and air-conditioning equipment, that are assembled to perform a function. A *component* is an item of mechanical or electrical equipment, such as a canister transfer machine, transport and emplacement vehicle, pump, valve, or relay.

thermal mechanical: Of or pertaining to changes in mechanical properties from effects of changes in temperature.

total system performance assessment: A *risk assessment* that quantitatively estimates how the potential Yucca Mountain repository system will perform in the future under the influence of specific *features, events*, and *processes*, incorporating *uncertainty* in the models and *uncertainty* and *variability* of the data.

transpiration: The removal of water from the ground by vegetation (roots).

transport: A process that allows substances such as contaminants, *radionuclides*, or *colloids*, to be carried in a fluid from one location to another. Transport processes include the physical mechanisms of *advection*, convection, *diffusion*, and *dispersion* and are influenced by the chemical mechanisms of *sorption*, leaching, precipitation, *dissolution*, and complexation. Also used to describe the movement of *magma* by flow within the Earth's interior.

tuff: A general term for volcanic rocks that formed from rock fragments and *magma* that erupted from a volcanic vent, flowed away from the vent as a suspension of solids and hot gases, or fell from the eruption cloud, and consolidated at the location of deposition. Tuff is the most abundant type of rock at the Yucca Mountain site.

uncertainty: How much a calculated or measured value varies from the unknown true value.

unsaturated zone flow: The movement of water in the unsaturated zone driven by capillary, viscous, gravitational, inertial, and evaporative forces.

variability (statistical): A measure of how a quantity varies over time or space.

NRC FORM 335
(12-2010)
NRCMD 3.7

U.S. NUCLEAR REGULATORY COMMISSION

BIBLIOGRAPHIC DATA SHEET

(See instructions on the reverse)

1. REPORT NUMBER
(Assigned by NRC, Add Vol., Supp., Rev., and Addendum Numbers, if any.)

NUREG-2109

2. TITLE AND SUBTITLE

Technical Evaluation Report on the Content of the U.S. Department of Energy's Yucca Mountain Repository License Application; Administrative and Programmatic Volume

3. DATE REPORT PUBLISHED

MONTH	YEAR
09	2011

4. FIN OR GRANT NUMBER

5. AUTHOR(S)

6. TYPE OF REPORT

Technical

7. PERIOD COVERED (Inclusive Dates)

8. PERFORMING ORGANIZATION - NAME AND ADDRESS (If NRC, provide Division, Office or Region, U. S. Nuclear Regulatory Commission, and mailing address; if contractor, provide name and mailing address.)

Division of High-Level Waste Repository Safety
Office of Nuclear Material Safety and Safeguards
U.S. Nuclear Regulatory Commission
Washington, DC 20555 0001

9. SPONSORING ORGANIZATION - NAME AND ADDRESS (If NRC, type "Same as above", if contractor, provide NRC Division, Office or Region, U. S. Nuclear Regulatory Commission, and mailing address.)

same as above

10. SUPPLEMENTARY NOTES

11. ABSTRACT (200 words or less)

This "Technical Evaluation Report on the Content of the U.S. Department of Energy's Yucca Mountain License Application; Administrative and Programmatic Volume" (TER Administrative and Programmatic Volume) presents information on the NRC staff's review of the U.S. Department of Energy (DOE) Safety Analysis Report (SAR), provided on June 3, 2008, as updated on February 19, 2009. The NRC staff also reviewed information DOE provided in response to NRC staff requests for additional information and other information that DOE provided related to the SAR. In particular, this report provides information on the NRC staff's evaluation of DOE's proposed administrative and programmatic activities regarding the following: Research and Development Program to resolve safety questions; Performance Confirmation Program; Quality Assurance Program; Records, reports, tests, and inspections; DOE organizational structure; Key positions assigned responsibility for safety and operations; Personnel qualifications and training; Plans for startup activities and testing; Plans for conduct of normal activities; Emergency planning; Controls to restrict access and regulate land uses; Uses of geologic repository operations area for purposes other than disposal of radioactive wastes

12. KEY WORDS/DESCRIPTORS (List words or phrases that will assist researchers in locating the report.)

10 CFR Part 63, Yucca Mountain, geologic repository, technical evaluation report, TER, U.S. Department of Energy

13. AVAILABILITY STATEMENT

unlimited

14. SECURITY CLASSIFICATION

(This Page)

unclassified

(This Report)

unclassified

15. NUMBER OF PAGES

16. PRICE

UNITED STATES
NUCLEAR REGULATORY COMMISSION
WASHINGTON, DC 20555-0001

———————

OFFICIAL BUSINESS

NUREG-2109

Technical Evaluation Report on the Content of the U.S. Department of Energy's Yucca Mountain Repository License Application - Administrative and Programmatic Volume

September 2011